# Clinical Nutrit[ion] for the Health Scientist

Author

**Daphne A. Roe, M.D.**
Professor of Nutrition
Division of Nutritional Sciences
Cornell University
Ithaca, New York

**CRC Press**
Taylor & Francis Group
Boca Raton London New York

CRC Press is an imprint of the
Taylor & Francis Group, an **informa** business

First published 1979 by CRC Press
Taylor & Francis Group
6000 Broken Sound Parkway NW, Suite 300
Boca Raton, FL 33487-2742

Reissued 2018 by CRC Press

© 1979 by Taylor & Francis
CRC Press is an imprint of Taylor & Francis Group, an Informa business

No claim to original U.S. Government works

A Library of Congress record exists under LC control number: 78031447

Publisher's Note
The publisher has gone to great lengths to ensure the quality of this reprint but points out that some imperfections in the original copies may be apparent.

Disclaimer
The publisher has made every effort to trace copyright holders and welcomes correspondence from those they have been unable to contact.

ISBN 13: 978-1-138-50606-0 (hbk)
ISBN 13: 978-1-138-55785-7 (pbk)
ISBN 13: 978-1-315-15041-3 (ebk)

Visit the Taylor & Francis Web site at http://www.taylorandfrancis.com and the
CRC Press Web site at http://www.crcpress.com

# PREFACE

Nutrition manuals and textbooks are usually written to meet the needs of nutritionists or would-be nutritionists. This manual has been written to meet the needs of health professionals who are not nutritionists, including physicians, nurses, and pharmacists. The aim is to acquaint readers in these professions with the principles of nutrition and the application of nutrition to the better delivery of health care.

Decision to arrange the contents of the manual under particular section headings was made to facilitate the readers' access to areas of nutrition information required in their daily patient or client contact. In Section I, the aim is to present information on the determinants of food choice and food intake. In order to be able to understand what a person will eat, when they will eat, and how much they will eat, it is necessary to understand the interrelationships of cultural, social, personal, situational, and medical factors which condition eating or drinking.

Section II is a compendium of basic nutrition. Inclusion of tables of the nutrient composition of foods and formulas is justified because of the frequently expressed need by health professionals for a ready source of reference for this information. Guidelines for nutritional assessment are included here in the hope that such assessments will become essential components of all health evaluations.

Section III addresses the widespread problems of food abuse in our society. Many chronic health problems have been linked to dietary excesses and/or to irregular or unwise eating habits. Within this general context, selection of topics has been dictated by the prevalence of specific food-related problems in community, clinic, and hospital practice.

Section IV relates to persons and patients at nutritional risk. Particular emphasis has been placed on the risks and outcome of prenatal malnutrition and the causes and effects of malnutrition in special age groups, particularly in the elderly, alcoholics, and in hospitalized patients.

Section V reviews newer information on drug-food and drug-nutrient interactions and incompatibilities. It is assumed that physicians, nurses, and pharmacists need to know the effects of diet and nutritional status on drug absorption, therapeutic response, and risk of adverse side-effects. An introduction to the subject of drug-induced nutritional deficiencies is included, particularly because of the incidence of such deficiencies among patients requiring long-term drug therapies.

Section VI outlines specific roles of different health professionals in offering nutritional care. Recommendations are given for means of developing and integrating nutrition services at the community and hospital level by a team approach, which requires collaboration of medical, nursing, and pharmacy personnel.

References are selective and have been deliberately held to a minimum required to support the text. The author is cognizant of the fact that new nutrition information and misinformation is constantly appearing in the lay press and in technical journals. Readers are therefore encouraged to seek professional assistance from consulting nutritionists who take primary responsibility for nutritional care and are better able to interpret nutrition reports as they appear in the press or via the media.

Daphne A. Roe

# THE AUTHOR

**Daphne A. Roe, M.D.**, is Professor, Division of Nutritional Sciences, Cornell University, Ithaca, New York and Adjunct Professor, Upstate Medical Center, State University of New York at Syracuse. She is also Consultant Dermatologist to the University Health Services at Cornell.

Dr. Roe obtained the M.B.B.S. degree from London University in 1946 and the M.D. degree from London University in 1950. Her postgraduate training was at the Institute of Dermatology in London, where she was awarded the Chesterfield Medal in 1950. In the period 1953—61 she held research appointments at the Massachusetts General Hospital and at the University of Pennsylvania.

Dr. Roe was appointed to the Cornell Faculty in 1961, where she has been actively involved in teaching and research in nutrition. Her current research is in the areas of drug-nutrient interactions and biochemical assessment of nutritional status.

She is a Member of the American Institute of Nutrition, the American Academy of Dermatology, the Society for Investigative Dermatology, and the Royal College of Physicians. She is a Fellow of the Royal Society of Medicine of London. In 1978 she was certified Specialist in the Human Nutritional Sciences by the American Board of Nutrition.

Dr. Roe has published more than 80 papers on nutrition and dermatology. She is also the author of three other books: *A Plague of Corn: The Social History of Pellagra*; *Drug-Induced Nutritional Deficiencies*; and *Alcohol and the Diet.*

# ACKNOWLEDGMENTS

I wish to acknowledge the assistance of the faculty of the Division of Nutritional Sciences at Cornell University in helping me to prepare this book. Special thanks are due to my secretary, Mrs. Beverly Hastings, for preparation, typing, and collation of the text. I would like to express my appreciation to my husband, Professor Albert Roe, for his unfailing support and for his guidance to me in efforts to avoid technical jargon.

# TABLE OF CONTENTS

# SECTION I
# DETERMINANTS OF FOOD INTAKE

## A. Why Foods are Selected

Men and women choose food and not nutrients. This statement can only be qualified insofar that those with nutrition knowledge or with preoccupation about certain nutrients may deliberately select food for their nutrient quality. Food choice is determined in both sexes and in all age groups, beyond infancy, by multiple factors which can be classified under these several headings: cultural, social, personal, situational, and medical.

### 1. Cultural

In the population, influence on food choice is through family tradition, which is to a greater or lesser extent determined by country of origin, by ethnicity, and by religious or traditional belief. Foods that are preferred are those that are familiar, or foods that were family foods in early life. Cultural factors which distinguish food selection from food avoidance may not be rational but stem from beliefs passed on from one generation to another. Particular foods may be considered to be male (appropriate to men) or female (appropriate to women). Food choice may be determined by season, whether the season is religious or climatic (fish is eaten during Lent and soup during winter). Prestigious foods are selected if income permits (steak is preferred or in some societal groups, white bread). Foods or natural food components may be considered to be magic, or to influence destiny, which even today influences the use of garlic. Certain foods are considered appropriate or inappropriate for particular meals (oatmeal for breakfast; no milk for dinner for those of French or Mediterranean origin). Food taboos exist for all of us — otherwise we would eat cats and beetles. Food avoidance may be on the basis of religious tenets (no pork for Jews and Mohammedans); country of origin and ethnicity influence food selection (tripe is pleasant to French people, and rice is preferred over bread by Chinese). Folklore tradition may determine the choice of food for invalids (chicken broth for those who are debilitated).

### 2. Social

Social influences on food choice include family likes and dislikes. Parental eating practices may be important (the daughter may not choose to eat green vegetables if this has not been the habit of the mother). Upward social mobility may determine preference for butter vs. margarine, and veal cutlet vs. baked beans. Peer group pressures are currently important and account for present-day popularity of yogurt and bagels. Foods may be chosen because they have an association with festive occasions (turkey); fads are likely to condition food choice and food avoidance (brown rice selected, convenience foods rejected because of "chemicals they contain").

### 3. Personal

Personal influences on food choice or rejection include early association (with comfort or relief of hunger — hence, a yearning by some for fish gravy or corn meal mush, which were given to them during their childhood in the Depression). Taste satisfaction may determine selection of salted nuts. Sensory pleasure conditions choice of ice cream sundaes. A sense of creativity determines a preference for home-baked bread. Emotional association with reward, happiness, or conversely, despair may partly explain choice of cookies (good boy or girl!) and avoidance of rice pudding ("starving people would be glad to eat it").

## 4. Situational

Food choice is restricted by poverty or a lack of transportation, by restricted availability of food in the store, by lack of food storage facilities, lack of cooking facilities, lack of skills in food preparation, and by lack of time to cook. Food choice may be determined by menu, whether this is in an extended care facility or an elaborate restaurant.

Merchandising, advertising, and food display are used extensively by the food industry to influence food selection, and they have been most successful in their goals. We would like to think that food labeling influences food selection, and for some this may be true, though the strategies of the food industry to influence food selection are more pervasive and persuasive. Food stamps influence food selection among those who are eligible, since certain foods are covered and others of a luxury nature cannot be purchased with food stamps.

## 5. Medical

When unpleasant symptoms are associated with the intake of specific foods or beverages, these tend to be avoided, or conversely, foods that are selected are those that are tolerated. For example, if milk, cereal products, and fats induce diarrhea, then food selection will be from among other types of food. Appetite or anorexia conditions food selection. Early satiety may be a major cause of reduced food intake in the postgastrectomy patient, though here the appetite reduction and food finickiness may be conditioned by the dumping syndrome. Taste loss or perversions of taste and smell may determine avoidance of some foods and preference for others. The cancer patient with taste perversion may perceive meat as tasting rotten, and therefore may be repelled by the sight of meat. Most importantly, iatrogenic diets limit food choice, and if these are used additively by a patient, food choice may be limited to a very few items. Self-prescribed diets are also important as a causative factor in the determination of a monotonous diet. The patient with anorexia nervosa will deliberately avoid carbohydrate-containing foods, through a belief which is associated with this weight-phobia disease.

In summary, preferences are for foods used as rewards, for prestige foods, for foods considered to have a special property, for foods believed to be tasty, for foods having sensory appeal, for foods that are filling, for cheap foods, sweet foods, foods that you can't get at home, for foods that are readily prepared or available, for foods promoted by the media, for foods that are familiar, for ethnic or regional foods, and for foods that are easy to chew. Dislikes are for foods enforced by parents, for unfamiliar foods, for foods inducing an adverse response, for foods that have been badly prepared, for foods difficult to cook or prepare, for substances not considered to be edible, for bland foods, for foods that are not filling or satisfying, for foods having peculiar smell, taste, or consistency, for foods that are difficult to chew, and for foods that are considered to be "good for you."[1]

## B. Constraints on the Intake of an Adequate Diet

An individual's ability to obtain an adequate diet may be limited by socioeconomic, educational, situational, or medical factors.

### 1. Socioeconomic and Situational Factors

Under conditions of poverty or when low-income individuals are unwilling or unable to obtain financial assistance in food purchasing (e.g., food stamps), high cost may limit the diet from the qualitative as well as the quantitative standpoint. In the U.S., lack of money is rarely the sole determinant of dietary inadequacy, but economic factors do determine the number of different foods consumed and hence, the variety of the diet. Many different societal groups consume diets which are inappropriate to their nutritional requirements. Dietary excesses are as much a problem and probably a greater problem than deficiencies. The detary constraints of elderly housebound people, indigent alcoholics, or pregnant adolescents, for example, are not primarily due to their common lack of income but rather to other factors which affect the environment in which they live.[2-5]

Opportunities for food selection can never be greater than the variety of food available. In institutions, where residents or patients have no access to outside sources of food, selection can only be from among food items provided by the facility on that day, week, or month. Similarly, if an individual family lives in a locality where there is one store and that one store only contains a narrow range of foods, it is from among these foods only that choice can be made. Similarly, those who have no home-cooking facilities and have to eat cheaply in restaurants are limited by the foods on the menu which they can afford.

### 2. Education

Lack of education is a major factor determining poor diets in low-income populations. Women who have had an 11th-grade education or less are unlikely to understand classification of foods according to food groups or nutrients supplied. They are more likely to divide foods according to like, dislike, and "will not eat," or according to the meal at which they think the food should be served; or they may consider foods as being fattening or nonfattening. Lack of knowledge about food is not explained by an absence of formal nutrition education. Rather, food concepts tend to run parallel to general educational level. If food is not considered from the standpoint of nutrition, then the risk is that food selections will be based on preference, the influence of advertising, and appearance of meal items. The social isolate, who is also uneducated, is at particular nutritional risk because he/she has never learned to eat more than a very few common foods. Such men and women are prejudiced against trying new foods and cannot overcome this prejudice by contact with others who know how to eat better.

For those who are bored because they don't know how to spend their leisure time and for those who are unemployed or underemployed because they lack job skills, overconsumption of junk foods is a means of breaking the daily monotony.[6]

On the other hand, there are men and women of advanced educational background who select foods inadequate in key nutrients because they adhere to an exotic food philosophy.[7]

### 3. Medical

Medical constraints on dietary adequacy include the following:

**Psychiatric illnesses** — Psychotic depression, especially, leads to an indifference to food and self-neglect, and organic brain syndromes lead to loss of memory and/or loss of incentive to prepare food. Bizarre eating behavior may occur in schizophrenics, particularly those with paranoid schizophrenia who may believe that food presented to them is poisonous. Anorexia nervosa, which is essentially a weight phobia, leads to self-starvation of adolescent girls with this disease, and less commonly, adolescent boys or older persons.

**Alcohol and drug addiction** — Multiple factors contribute to deficient diets consumed by alcoholics. These include not only financial constraints on food purchase and situational factors which limit food preparation, but also associated depression, indifference, anorexia, inebriation, knowledge that desired psychological effects of alcohol can be more rapidly achieved in the fasting state, and adverse or unpleasant reactions following food consumption due to the effects of alcohol on the gastrointestinal tract (nausea, vomiting, diarrhea, and abdominal pain).[8]

In narcotic and other drug addicts grocery money may be diverted to purchase drugs. Anorexia may occur in drug addicts, more particularly in those who are consuming amphetamines. Intravenous injection of heroin can lead to serum hepatitis which causes anorexia and hence decreased food intake. Poor eating habits of drug addicts may also be due to general disinterest in food or to a rejection of conventional food-related practices.[9]

**Physical diseases causing anorexia** — Medical diseases including acute infections, chronic infections, and other chronic diseases including cancer may produce such a marked reduction in appetite that food intake is grossly reduced. Diseases particularly associated with anorexia are cancer and congestive heart failure. In cancer patients, anorexia may be associated with appetite perversion and particular aversion to meat and meat products. Congestive heart failure from various antecedent causes is associated both with anorexia and early satiety. Consumption of large quantities of foods by causing gastric distention can lead to cardiac embarrassment. The chief cause of malnutrition in postgastrectomy patients is failure to consume enough food due to early satiety and the induction of the dumping syndrome by certain foods including foods and beverages containing concentrated sugars.[10]

**Therapeutic drugs** — Therapeutic drugs can lead to a reduction in food intake because of direct or indirect anorectic effects. Anorectic drugs include those which impair taste, cause aversion to food by inducing nausea, vomiting, pain, or diarrhea, as well as drugs such as amphetamines used purposefully as appetite suppressants, because they induce central depression of appetite.[11]

**Iatrogenic diets** — Iatrogenic diets prescribed by physicians can cause malnutrition. Therapeutic diets which are notoriously liable to be nutritionally inadequate include stringent ulcer diets, elimination diets, and low-sodium diets. Unpalatable diets remain uneaten. Nutritional risks of therapeutic diets are highest when the patient uses dietary suggestions additively with exclusion of a wide variety of foods essential to dietary requirements.[12]

**Physical handicaps** — Chronic crippling diseases such as arthritis, late effects of stroke, and multiple sclerosis can limit food access not only because marketing and food preparations are difficult, but also because eating is a problem. Masticatory inefficiency due to the edentulous state or to ill-fitting dentures may induce elderly people not to eat meats and other foods that have to be chewed. Geriatric disability can cause dietary inadequacy because of mental as well as physical handicaps.[13]

## C. Food Fallacies

Folklore, ancient and modern, as well as the persuasive writings of those who create popular diets, influence health professionals and the lay public to believe nutrition nonsense. The following statements, which are in italics, are not true. Health professionals should explain to these patients why these statements are fallacies.

### 1. Starchy Foods (Potatoes, Bread) are Fattening

The fattening quality of food is only dependent on food energy consumed above the requirements of the body. Potatoes, bread, and other starchy foods have no special effect on body weight and indeed should form a part of the normal diet.

### 2. Grapefruit and Cottage Cheese are Slimming

There are no foods which make you lose weight. Bulky foods with a low-caloric density may induce a feeling of satiety so that less food is actually eaten.

### 3. Cheese is Constipating
There is no evidence that cheese causes constipation. Low-residue diets in general (i.e., diets which do not contain enough dietary fiber) are likely to cause constipation. Cheese is an excellent food.

### 4. Hot Food is Better for You than Cold Food
There is no scientific evidence that the nutritional value of food is improved by warming it. Only the sensory appeal of food can be improved by serving it hot.

### 5. Hypoglycemia (Low Blood Sugar) is Caused by Eating Too Much Sugar and Hypoglycemia Causes Irritability, Lack of Concentration, Behavioral Disturbances, Poor Performance in School, and General Lack of Pep
Normal blood sugar values are well maintained in people with different eating patterns provided that they are not diabetics, alcoholics, or have insulin-producing tumors of the pancreas. The symptoms, which have been ascribed to hypoglycemia, may be due to situational problems or psychological problems which require counseling.

### 6. Raw Vegetables Give You Gas
The only vegetables which give you gas are beans, and then only when very large quantities are eaten. It is most desirable that raw vegetables be part of your daily diet because they are major sources of vitamins including vitamin A in the form of carotene, folic acid, and vitamin C.

### 7. Vitamin C Prevents/Cures the Common Cold
Vitamin C does not prevent or cure the common cold. Extensive investigations which have been carried out giving normal subjects megadoses of vitamin C or dummy pills have not supported claims that vitamin C significantly affects the occurrence or course of the common cold.

### 8. Orange Juice is Too Acid for Infants and Young Children
Wrong! Orange juice is well tolerated by infants from the age of 3 months and is an excellent source of vitamin C, which is deficient in milk formulae.

### 9. Adolescent Acne is Caused by Eating Too Much Chocolate, Peanut Butter, Cheese, Potato Chips, Nuts, and French Fried Potatoes
Whether or not a boy or girl has severe acne depends on the parents they chose (familial tendency) and on precise hormone balance. The foods cited are foods most kids enjoy, and many parents disapprove of, but these foods do not cause acne!

### 10. Vitamins, If Taken Regularly, will Prevent Liver Disease from Alcohol Abuse
Alcohol damages the liver and the degree of damage is related to the level and duration of alcohol intake. No vitamin or combination of vitamins will prevent alcoholic liver disease in an alcoholic.

### 11. Natural Vitamins and Minerals from Natural Sources (e.g., Iodine from Kelp and Calcium from Bone Meal) have Beneficial Effects on Health Which Cannot be Obtained from Single or Multiple Vitamin-Mineral Preparations Manufactured by Pharmaceutical Companies
Natural vitamins have no special properties or health-promoting effects which are different from regular vitamins. Similarly, there is no advantage to be gained from taking kelp or bone meal because iodine and calcium can better be obtained from

iodized salt and milk, respectively. There are risks in taking kelp which contains arsenic and bone meal which may be a source of lead.

### 12. Food Additives Cause Hyperactivity and Related Behavioral Problems in Children

Controlled experiments have not supported the idea that additives currently in food or in beverages cause or influence hyperactivity.

### D. Smoking and Its Effects on Eating Habits

Eating habits of smokers may differ from those of nonsmokers. Changes in eating habits, associated with smoking, are in part attributable to the depressant effect of nicotine and/or volatiles inhaled with cigarette smoke on appetite, and are due in part to personality traits or personal habits of the smoker. Studies supporting an anorectic effect of smoking must be interpreted with caution. The finding that weight gain commonly follows discontinuation of smoking could reflect various changes that take place. There is an increased intake of food when smoking is no longer practiced because of a desire to relieve tension by eating, or because local irritant effects of smoking on the upper gastrointestinal tract are removed, or because systemic toxicity from nicotine is obviated. Another consideration is that taste and smell sensitivity are reduced by heavy smoking and will return to normal when smoking is stopped. The lay observation that food "tastes" better and is therefore more appealing when smoking is stopped is based on sound scientific evidence.

Further, there is evidence from British studies that the nonsmokers' and perhaps the ex-smokers' attitudes to food differ from those of the smoker. The nonsmoker not only refuses to smoke on the grounds of the health risks of smoking, but also is more likely to select foods on the basis that they promote good nutrition. Differences are particularly to be found among vegetarians who seldom smoke, but who choose their vegetarian diet carefully for nutrient content.

Alcoholics, on the other hand, are commonly smokers, and often heavy smokers, and their self-abusive habits may extend to food. Studies of the food habits of alcoholics by the author have shown that intake of vegetables is significantly lower than in matched groups of low-income nonalcoholics, and that the heavy smokers among the alcoholics fall into the low or nonvegetable eating group. From these studies, there can be no implication that smoking conditions the eating of vegetables directly, and explanations for avoidance of vegetables could relate to economic factors or situational factors, whereby the purchase or preparation of vegetables is more difficult for the alcoholic, especially one who does not have a stable home.

The assumption that nonsmokers are more likely to seek and choose food for good health must be tempered by a general observation that nonsmokers and ex-smokers tend to eat more candy than smokers.

In pregnant women the average birth weight of the infants of smokers is lower than that of nonsmokers. Whereas this difference may be explained by some investigators as a direct toxic effect of smoking on fetal development, there is also information that food intake of pregnant smokers is lower than that of nonsmokers.[14,15]

# SECTION II
# BASIC AND APPLIED CONCEPTS OF HUMAN NUTRITION

## A. Definitions of Food Energy and Nutrients
### 1. Basic Concepts of Human Nutrition

People have defined energy and nutrient requirements for growth, biochemical maintenance, and reproduction. In this context, energy refers to the release of power from food fuels to drive the vital processes of the cell or body. In the human body, food energy is derived from the breakdown of carbohydrates, fats, and proteins. Within the body utilization of energy for physical work or movement and for anabolic (build-up) activities can only be derived from that portion of the total food energy consumed which can be converted into high-energy compounds — such as adenosine triphosphate (ATP) — which are the actual releasers of productive energy. Metabolic rate refers to the rate at which the body consumes energy.[16]

Nutrients are those chemicals, derived from food or otherwise from the physiological environment, which when absorbed, perform essential functions. Nutrients are classified in several ways. Macronutrients include carbohydrates, fats, and proteins which are not only ready sources of food energy but which are also variously utilized for fuel storage, synthesis of new tissue, production of catalysts or enzymes, and in the maintenance of the internal milieu which is called homeostasis. Micronutrients include vitamins and minerals which promote and control metabolism of macronutrients, participate in enzyme function, are required for the integrity of tissues, are necessary to the synthesis of key compounds such as hemoglobin needed for oxygen transport, and with macronutrients are required in the defense mechanisms of the body. Essential nutrients are those which cannot be synthesized within the body and must therefore be derived from food. Nonessential nutrients can, on the other hand, be synthesized in the body, but may also be obtained from the food supply. Whether or not a nutrient is essential is dependent on genetic and acquired factors.

## B. Energy Requirements

Energy requirements depend on: physical activity, body size and composition, age, and the environmental temperature. For persons of the same age and size, living in the same environment and performing the same amount of physical work, energy requirements will be similar. There is, however, reason to believe that there may be interindividual differences in the efficiency of food-energy utilization including variability in digestion and absorption of essential energy sources. For body weight to be maintained, energy intake must balance energy expenditure. When energy balance is maintained, intake must equal energy output, plus energy gain. Some increase in metabolic rate occurs following intake of foods, and such a physiological increase in the metabolic rate is absent in the fasting individual.

The body requires energy for cellular activities in which more complex components are produced from simple components, for maintenance of body temperature, and for performance of involuntary as well as voluntary movements. The traditional unit of energy is the calorie, which is the amount of heat required to raise 1 g of water through 1°C. But for convenience, food energy has been and still is measured in kilocalories (kcal) which is the common unit of food energy referred to in food composition tables and diet manuals. In the International System of Units, accepted in 1960, the calorie is replaced by the joule, a unit which can be used interchangeably for different forms of energy — 1 cal is equivalent to 4.184 (4.2) joules. Larger units based on the joule are the kilojoule (kJ) and the megajoule (MJ), which is equivalent to 1000 kJ.

The amounts of energy resulting from the oxidation of carbohydrates, fats, and proteins have been determined in the laboratory by the use of a bomb calorimeter. We know that in the body using direct or indirect calorimetry, the following figures for energy production from these foods can be generalized:[17] 1 g carbohydrate supplies 4 kcal (16 kJ); 1 g fat supplies 9 kcal (37 kJ); and 1 g protein supplies 4 kcal (17 kJ).

## C. Nutrients, Their Function and Their Food Sources
### 1. Carbohydrates
Food carbohydrates consist of two major classes, starches and sugars. Starches, which are polymers of the simple sugar glucose, are found in root vegetables, tubers, legumes, seeds, and many fruits. Sugars, of importance in human nutrition, are the disaccharides: sucrose, maltose, and lactose; and the monosaccharides: glucose, fructose, and galactose. Sucrose is the sugar in ordinary table sugar, derived from sugarcane or sugar beet. Sucrose can be split by the process of hydrolysis to its component monosaccharide sugars which are glucose and fructose. Maltose, which occurs in malt liquors and in starch hydrolyzates, is composed of two glucose units. Lactose, or milk sugar, consists in one unit of glucose and one unit of galactose. Lactose is a component of baking mixes and certain infant foods, as well as all foods to which milk or dried milk powder are added. Glucose, which is the chief sugar in the blood of most animals including the human, is found in the free state in ripe fruits, roots, leaves, and in the flowers and sap of plants. Dextrose and corn syrup are commercial glucose products but may contain other carbohydrates. Fructose is present in honey, syrup, molasses, and ripe fruits. Galactose is present in the combined form in all foods that contain lactose and may be present in certain fruits such as grapes and in legumes. Glycogen, the principle storage form of carbohydrate in animals, is particularly rich in liver. It is a polymer consisting of repeating glucose units.

Carbohydrate foods constitute about 50% of daily energy intake in the U.S. In recent years intake of complex carbohydrates from starchy foods has been reduced and at the same time sugar intake has increased, more especially in soft drinks, convenience foods, and in sweets. Carbohydrates are an excellent source of food energy and are needed in order that the body can metabolize fat efficiently. Further, adequate intake of carbohydrates has a sparing effect on protein needs.

Excessive intake of soluble sugars, particularly sucrose, has etiological significance in the development of dental caries and may be a factor explaining the increasing incidence of maturity onset diabetes in populations that have changed from a low-sucrose to a high-sucrose diet.

### 2. Fats and Other Lipids
Lipids are defined as naturally occurring substances which are insoluble in water but soluble in organic solvents. Important dietary lipids are triglycerides, which are esters of glycerol and fatty acids, forming fat or oils, sterols such as cholesterol and ergosterol (vitamin $D_2$), and the other fat-soluble vitamins A, E, and K.

Food fats vary in their fatty acid composition. Fatty acids are classified as saturated or unsaturated. The most common saturated fatty acids in animal fats are palmitic acid and stearic acid. Foods containing saturated fat (high in saturated fatty acids) are beef, lamb, pork, veal, whole milk, cheeses, cream, hydrogenated shortenings, butter, some margarines, ice cream, chocolate, and coconut oils. Unsaturated fatty acids are found in fats of animal and plant origin. Unsaturated fatty acids in animal fats are oleic acid and palmitooleic acid. Linoleic and linolenic acids are major unsaturated fatty acids in plant oils. Animal foods containing monounsaturated fat are chicken, duck, and turkey. Foods containing polyunsaturated fatty acids include fish and also soybean, cottonseed, safflower, and sunflower oils. Major food sources of monounsaturated fatty acids are vegetable oils including olive oil, soybean oil, peanut oil, and cottonseed oil.

In the U.S. the fat content of the average diet is presently more than 41% or about 160 g per day per person. Two thirds of the fat in the American diet is from animal sources, especially sources of saturated fats, and the other one third is from vegetable sources including oils and hydrogenated fats or margarines. Dietary fat fulfills the following functions: it serves as an important energy source; it provides a source of essential fatty acids; and it acts as a carrier for fat-soluble vitamins. Fat requirements could be met by a diet containing not more than 25 g per day per person of food fat.

Essential fatty acids are linoleic acid and linolenic acid, which cannot be synthesized in mammalian tissues, and arachidonic acid, which is formed from linoleic acid. Essential fatty acids are required in the synthesis of the prostaglandins. Signs of essential fatty acid deficiency include dermatitis, impaired caloric efficiency, decreased resistance to physical stress, as well as an impairment in lipid transport. Essential fatty acid deficiency may occur in patients on total parenteral alimentation without an added fat source or source of essential fatty acid.

Cholesterol is the commonest sterol in animal tissues. Like other sterols, it is a higher molecular weight alcohol with a multiple ring structure. Major food sources of cholesterol are: eggs, liver, meats, and meat products including sausage, cold cuts, and frankfurters, as well as shellfish and cheese. Functions of cholesterol are in the synthesis of hormones and in the preservation of cell integrity. It is also a precursor of the intermediate 7-dehydrocholesterol, which is a precursor of vitamin $D_3$. The American diet has a high cholesterol content due to large consumption of eggs and meats. High intake of cholesterol is associatd with increases in blood cholesterol. Epidemiological surveys indicate association between cholesterol intake and the incidence of atherosclerotic heart disease, as well as association between high blood cholesterol and heart disease. Diets high in saturated fat also tend to raise serum cholesterol.

## 3. Proteins and Amino Acids

Proteins are natural substances of large molecular weight which always contain nitrogen, carbon, hydrogen, and oxygen and may also contain sulfur, phosphorus, and metallic elements such as zinc, iron, and copper. Protein sources in the diet include animal and vegetable foods. Following digestion or hydrolysis, proteins yield smaller molecular weight compounds called amino acids. Amino acids are divided into two groups, essential and nonessential. Essential amino acids are those that cannot be synthesized in the body. The essential and nonessential amino acids are listed in Table 1. Major food sources of proteins and hence, of amino acids are meats, fish, milk, cheese, eggs, legumes, and to a lesser extent, cereal grains. When protein foods are eaten, they are digested and broken down to peptides and amino acids which are absorbed and from which tissue proteins are formed. The nutritional value of a food protein depends on the relationship of its amino acids and those which are required for the synthesis of body proteins. If the diet is deficient in essential amino acids, protein status cannot be maintained and wasting occurs. Food sources which lack adequate amounts of essential amino acids include proteins in corn, wheat, and legumes which lack tryptophan, lysine, and methionine, respectively. The proper balance of essential amino acids can be obtained by combining these food sources of protein in the diet in the proper proportions. The need for such food combinations is to support good nutrition in those who adhere to a vegetarian diet containing no animal proteins (vegan diet).

Dietary protein is required for growth, tissue maintenance and repair, and synthesis of enzyme proteins, nutrient transport proteins, and proteins required for the immune processes or defense mechanisms of the body.

TABLE 1

Essential and Nonessential Amino Acids Required
by People for the Biosynthesis of Proteins

Essential Amino Acids

| | |
|---|---|
| Isoleucine | Phenylalanine |
| Leucine | Threonine |
| Lysine | Tryptophan |
| Methionine | Valine |
| Histidine (infants and young children) | |

Nonessential Amino Acids

| | |
|---|---|
| Alanine | Histidine (normal adolescents and adults) |
| Arginine | Proline |
| Aspartic acid | Serine |
| Cysteine | Tyrosine |
| Glutamic acid | Asparagine |
| Glycine | Glutamine |

When mixtures of essential and nonessential amino acids are given orally or intravenously under conditions where food proteins cannot be digested, it is essential that amino acid imbalance be avoided. If amino acid imbalance occurs, protein synthesis will be impaired.

In parenteral nutrition a ratio of essential to total amino acids of 0.20 has been accepted for adults and of 0.40 for infants. More recent information suggests that the ratio of essential to total amino acids should be 0.40 for all age groups except for those whose liver or kidney function is impaired. L-amino acids are better utilized than DL-amino acids. Histidine may be an essential amino acid in uremic adults as well as in growing infants. In intravenous (parenteral) nutrition, arginine may be required to prevent hyperammonemia. Proline is required for collagen synthesis necessary for wound healing. Alanine is necessary for the transport of amino nitrogen between tissues. Health professionals involved in the nutritional support of alcoholics, renal patients, and patients with metabolic disorders should be aware that amino acid requirements change with impairment of liver and kidney function as well as with the integrity or otherwise of enzyme systems needed for normal amino acid utilization.

## 4. Vitamins

Vitamins are a heterogeneous group of nutrients which are required in the diet for essential metabolic functions, structural integrity, and physiological adaptation. Vitamins are divided into the water-soluble vitamins and the fat-soluble vitamins. The water-soluble vitamins consist of the B vitamin complex and vitamin C. B vitamins which are essential to life, and which in human beings must be obtained from the diet, are thiamin, niacin, riboflavin, folacin, pyridoxine (vitamin $B_6$), cyanocobalamin, and vitamin $B_{12}$. Other B vitamins are pantothenic acid, which is widely distributed in food, and biotin, which is not only available from the diet but also through synthesis by the intestinal microorganisms. In the body, B vitamins are converted to their active forms, which function as coenzymes in vitamin-dependent metabolic processes. The B vitamins have four major functions as follows: (1) they are energy-releasing agents, (2) they have hemopoietic functions, that is, they are necessary to the maturation of blood cells and the synthesis of hemoglobin, (3) they are necessary to the nervous system and the transmission of nerve impulses, and (4) they are necessary for the metabolism of foreign compounds. Functional classification of B vitamins shows that each may have several functions.

TABLE 2

Food Sources and Functions of Water-Soluble Vitamins

| Vitamin | Functions | Major food source |
|---|---|---|
| Thiamin | Required in carbohydrate metabolism and for integrity of brain and nerve function | Whole cereal grains; enriched breads and cereals; nuts; pork; eggs; milk and potatoes (if used extensively) |
| Niacin | As active coenzymes important in hydrogen transport, in carbohydrate, lipid, and protein metabolism | Liver; lean meats; fish and shellfish; enriched breads and cereals; eggs |
| Vitamin $B_6$ | Necessary for the metabolism of amino acids, the synthesis of neurotransmitters, and formation of heme as in hemoglobin | Liver; meats; fish; eggs; whole grain cereals; green leafy vegetables; bananas, grapes, and pears |
| Riboflavin | As flavin coenzymes necessary to hydrogen (electron) transport; involved in lipid metabolism, amino acid metabolism, activation of vitamin $B_6$, and drug metabolism | Milk; eggs; liver; lean meats; enriched bread and cereals; leafy green vegetables |
| Folacin | Active coenzyme forms are required in the synthesis of the nucleic acid DNA, in the maturation of red and white blood cells, and in the maintenance of epithelial integrity | Green leafy vegetables; liver; fortified breakfast cereals; yeast; legumes; citrus fruits |
| Vitamin $B_{12}$ | Necessary for DNA synthesis, maturation of red and white blood cells, and integrity of nerve fibers | Animal protein foods |
| Vitamin C | Major roles are in hydroxylation, in synthesis of normal connective tissue, and in drug metabolism | Citrus fruits and fruit juices; other fresh fruits; green peppers; tomatoes; cabbage; broccoli; peas and breakfast cereals; fruit beverages fortified with the vitamin |

Vitamin C acts in maintaining the structural integrity of connective tissue and acts as a reducing agent, promoting iron absorption from the intestine. It is necessary to drug metabolism.

In Table 2, water-soluble vitamins are classified functionally and the major food sources of these vitamins are indicated.

Four fat-soluble vitamins are recognized: vitamins A, D, E, and K. Vitamin A is obtained from the diet in two forms, namely as carotene, which is converted to vitamin A, and vitamin A itself. Vitamin A intake is a term used collectively to describe intake of both carotene and preformed vitamin A. Functions of vitamin A are (1) in the visual process, (2) in maintaining the structural integrity of epithelia, (3) in reproduction, and (4) in growth. Vitamin D, a sterol, is obtained in part from the diet and in part from synthesis in the skin under the action of ultraviolet light. The vitamin functions

TABLE 3

**Food Sources and Functions of Fat-Soluble Vitamins**

| Vitamin | Functions | Major food source |
|---------|-----------|-------------------|
| A | Required in the visual process, in membrane stability, and in spermatogenesis | *Preformed*<br>Milk and milk products, liver, kidney, eggs<br>*β-Carotene precursor*<br>Carrots, sweet potatoes, yellow corn, winter squash, green, leafy vegetables, pumpkin, tomatoes |
| D | Required for calcium absorption and normal bone formation | Liver and vitamins, D-fortified milk, butter and cereals |
| E | Lipid antioxidant; preserves the integrity of the red blood cell | Soybean oil, nuts, wheat germ oil, eggs, liver, margarines |
| K | Necessary to the synthesis of normal blood-clotting factors | Green, leafy vegetables |

are such that it is now considered that vitamin D might be classified as a hormone. Vitamin D is necessary to calcium absorption, calcium homeostasis, and bone growth.

Vitamin E functions as an antioxidant and in this capacity it prevents hemolysis (bursting) of red blood cells. Other metabolic functions of vitamin E which have been identified in experimental animals have not been well defined in humans.

Vitamin K is supplied both from the diet and also by synthesis due to the activity of the intestinal bacteria. It is important in the synthesis and hence, in the functioning of factors involved with blood clotting. Functions of fat-soluble vitamins and sources of these vitamins are summarized in Table 3.[18-20]

### 5. Minerals

Minerals can be divided into macronutrients and trace elements. Mineral macronutrients include sodium, chloride, potassium, calcium, phosphorus, magnesium, and sulfur. The definition macronutrient implies that these minerals are required in amounts in excess of 100 mg per day. Mineral micronutrients or trace elements include iron, copper, cobalt, zinc, manganese, iodine, molybdenum, selenium, fluorine, and chromium. Trace elements are needed at a level not exceeding a few milligrams per day.

### a. Sodium, Potassium, and Chloride

Sodium and potassium contribute significantly both to extracellular and intracellular osmolarity. Sodium is the major cation in the fluid which surrounds cells (interstitial fluid) and potassium is the predominant intracellular cation. Under equilibrium conditions, the effective osmotic pressures of these two compartments are balanced.

Sodium and potassium are involved in the electrophysiology of cells and in the propagation of nerve impulses. These elements play essential roles in acid-base balance. It is also clear that small changes in the extracellular concentration of sodium affects blood pressure and that small changes in the blood and tissue concentration of potassium alter cardiac function. Active transport of nutrients, particularly glucose across the intestinal mucosa, is sodium dependent. Protein and amino acid metabolism are sodium and potassium dependent. Insulin secretion is depressed by potassium depletion.

High-sodium intakes are generally undesirable, especially since in some people, dietary sodium raises blood pressure, and in hypertensives, restricted sodium intake is an important component of treatment. Chronic sodium chloride toxicity associated with excessive intake causes an increase in extracellular fluid volume and, conversely, restriction in sodium intake diminishes extracellular fluid. However, it seems clear that genetic differences exist with respect to sodium handling and sodium-induced hypertension. The relationships between increased extracellular fluid volume and the induction of hypertension are not well defined. Meneely and Battarbee, who have reviewed pertinent literature, conclude that in the U.S., extracellular fluid volumes above the normal range reflect excess dietary sodium.[21] At least in laboratory rats, increased potassium intake ameliorates chronic sodium toxicity. Current information indicates that a national dietary goal should be to decrease sodium and increase potassium intake.

High-sodium foods are bacon, ham, hot dogs, salami, chipped beef, olives, canned fish, potato chips, pretzels, other salted snack foods, and canned vegetables. High-potassium foods are oranges and orange juice, bananas, fresh vegetables such as peas and beans, and milk.

The public needs to be discouraged from adding salt or other sodium-containing flavoring agents such as monosodium glutamate in excessive amounts to foods. Infant foods should not be salted, since infants are at particular risk for developing sodium toxicity.

Chloride is the most important anion both in the maintenance of fluid and electrolyte balance. It is also needed to form hydrochloric acid in gastric juice. Chloride is mainly obtained from the diet as sodium chloride. If sodium-restricted diets are prescribed, then alternate sources of chloride may be required.[22]

### b. Calcium

Mineralized tissues in the human body consist of the bones and the teeth. Bone mineral consists of two forms of calcium phosphate which are deposited in an organic matrix made up of collagen, which is a major fibrous constituent of bone and other connective tissues, and mucopolysaccharides, which form a cohesive interfibrillar gel. The calcium salts in bone contribute to its hardness, and hence, affect structural support. The one percent of calcium which lies outside the skeleton includes the soluble forms of calcium, which are required for blood coagulation, skeletal muscle, contractility of heart muscle, nerve function, and activity of enzyme systems. The calcium content of the blood is held very constant because it is regulated by a number of hormones including the active form of vitamin D, calcitonin, and parathyroid hormone. Parathyroid hormone mobilizes calcium from bone when decreases in serum calcium occur. Parathyroid hormone also stimulates formation of the active form of vitamin D which promotes calcium absorption. Calcitonin release is in response to elevation in serum calcium, and this hormone promotes deposition of calcium in bone.

There has been much debate on dietary calcium requirements. Dietary calcium should be at a level to make up for losses via the urine and feces and to reduce the need for the body to utilize bone calcium to maintain required calcium levels in blood and tissues. In middle or later life, there may be an extensive loss of bone and bone mineral which is designated osteoporosis. Osteoporosis is the underlying cause of hip fractures and back problems occurring in older people.

The simple assumption that, by consuming enough food containing calcium, it is possible for men and women to prevent osteoporosis is not proven.

Present recommendations for calcium intakes are based on levels which in several studies have been shown to produce calcium balance. Calcium requirements are greater with high-protein diets. Absorption of calcium is lower from foods such as oatmeal containing phytate which binds the calcium in the intestine. Some evidence from animal and human studies points to a relationship between periodontal disease and inadequate calcium intake over time.

High-calcium foods are milk, cheese, and yogurt. Patients with milk intolerance, including milk allergy and lactose intolerance, should be advised that in order to meet their calcium needs, calcium lactate tablets should be required. Bone meal is not recommended as a calcium source because of current risks of lead contamination.[23]

### c. Phosphate

Dietary phosphorus (phosphate) is required in the formation of bone phosphate and to supply cellular phosphorus needs. Requirements are higher during the period of active growth. The actual amount of dietary phosphorus needed in childhood or in adult life is dependent on the availability of phosphorus from the foods consumed and on urinary losses of phosphate. Phosphorus absorption is also dependent on vitamin D status. Phosphate depletion occurs if phosphate reabsorption in the kidney is impaired, if availability of dietary phosphorus is low, and if antacids are taken to excess.

High-phosphorus foods are those of animal origin. In the U.S., due to the high consumption of rich phosphorus sources such as meat by people whose calcium consumption is low, the ratio of calcium to phosphorus in the diet may be unfavorable and possibly may contribute to the development of osteoporosis.[24]

### d. Magnesium

Magnesium is necessary for neuromuscular transmission. Excellent sources of magnesium are legumes, including both beans and peas, potatoes, other foods of vegetable origin, and dried fruits. Deficiency is associated with impaired utilization of thiamin (vitamin $B_1$) and variably with evidence of both calcium and potassium depletion. Signs of deficiency include tremor, tetanic convulsions, and coma if the deficiency is not rapidly reversed. Alcoholics may develop magnesium deficiency due to low intake and excessive urinary loss.

Symptoms of delirium tremens and other withdrawal states in alcoholics have been attributed to magnesium deficiency. However, neither delirium tremens nor other neurological signs occurring during alcohol withdrawal respond predictably to magnesium administration.[25,26]

### e. Sulfate

Most of the sulfur in the human body is in the form of sulfur-containing amino acids including methionine, cysteine, and cystine. The sulfur in proteins is present in the form of sulfur-containing amino acids.

Sulfur-containing amino acids in the diet are metabolized to yield sulfate which is utilized in the formation of sulfoconjugates, some of which have biological functions but many of which are excreted in the urine. Sulfates in the urine include inorganic sulfates and organic sulfates. Organic sulfates are metabolites or soluble excretion products of endogenous compounds such as steroids and foreign compounds such as drugs. As far as is known in the human, there is no requirement for sulfate in the diet because the sulfate used in the synthesis of retained and excreted inorganic and organic sulfates is provided by sulfur-containing amino acids in dietary proteins.[27]

### f. Iron

Iron is an essential nutrient. It is needed for the formation of hemoglobin and other heme compounds which fulfill important physiological functions in oxygen transport and tissue respiration. Iron is present in the diet in two major forms, heme iron and nonheme iron. Heme iron is the iron present in animal protein foods such as meats, fish, and eggs. Nonheme iron is present in foods of vegetable origin. Bread and cereals are fortified with inorganic iron. Absorption of heme iron is better than nonheme iron. Iron absorption is also highly influenced by the presence of other nutrients and non-nutrients in the diet. Vitamin C promotes iron absorption. Iron absorption is decreased by eating cereal foods containing bran and by drinking large amounts of strong tea. The total amount of iron in the diet is related to the food-energy intake such that 6 mg of iron is usually ingested for every 1000 kcal.

Severe iron deficiency causes anemia. Anemia occurs if iron intake and absorption is insufficient to meet bodily needs. Bodily needs for iron are high during the period of active growth and in women during the reproductive phase of life. Nonpregnant women need dietary iron to compensate for iron lost during menstruation. Pregnant women have high iron requirements to supply the needs of the developing fetus. Control or restriction of caloric intake is necessary whenever it may be important for women to take a small daily iron supplement to prevent iron deficiency. Alternate methods of increasing dietary iron are being explored, including higher fortification of staple foods such as bread and wheat flour.[28-31]

### g. Iodine

Iodine is required for the synthesis of iodine-containing thyroid hormones. Iodine deficiency is associated with the development of goiter, which used to be endemic in areas where iodine was deficient in the soil and water, and hence, in the animals, cereals, and vegetables used for food.

If a pregnant woman is severely iodine deficient, her baby may be a cretin, having a nonfunctional thyroid gland. Cretins are dwarfed and mentally retarded unless given thyroid hormones from early life.

Iodine in the American diet comes from a number of sources. Today the level of iodine intake is related to the amount of milk drunk, the amount of iodized salt consumed, and the amount of bread eaten which had potassium iodate added to the dough. Iodine intake also varies with the amount of iodine in the water supply and with the amount of food eaten which contains erythrosine, an iodine-containing food-coloring agent. Iodine in milk is in large part a contaminant coming from the iodine-containing antiseptics (iodophors) which are used to wash down the cows' udders.

Excess iodine may cause iodide goiter, which can be associated with the development of hyperthyroidism.[32,33]

### h. Zinc

Functions of zinc are as a constituent of enzymes which are concerned with the metabolism of nucleic acids, protein, carbohydrate, and alcohol. Zinc is both a structural and a functional component of the digestive enzyme carboxypeptidase. It is also present in the insulin molecule. Zinc is required for normal transport and metabolism of vitamin A. Essential roles in dietary zinc are related to pre- and postnatal growth, wound repair, keratinization (formation of the horny surface layer) of the epidermis, immune function, taste acuity, and visual function.

Rich food sources of zinc include meat, liver, eggs, and seafood. Milk is also a good source of available zinc. Whole grain cereals contain zinc, but zinc availability from these foods is limited.

Zinc deficiency associated with dwarfism and failure of sexual maturation was first identified in Iranian and Egyptian boys. Acrodermatitis enteropathica is an inherited

TABLE 4

Functions, Sources, and Deficiencies of Selected Trace Elements

| Trace element | Function | Source | Signs of deficiency |
|---|---|---|---|
| Chromium | Insulin cofactor | Yeast, liver, meats | Impaired glucose tolerance |
| Cobalt | Component of vitamin $B_{12}$ | Meats, liver | Megaloblastic anemia and nerve degeneration in vitamin $B_{12}$ deficiency |
| Manganese | Required for normal bone structure, reproduction, and neural function | Kidney, liver, sweetbread, nuts, cereals, tea | Impaired growth, skeletal abnormalities, impaired reproduction, ataxia in animals, unknown in humans |
| Selenium | Essential component of enzyme glutathione peroxidase | Liver, kidney, fish | Liver necrosis and muscle disease in animals — unknown in humans |
| Molybdenum | Protects dental enamel | Cereals, legumes, kidney, liver | Dental caries? |
| Fluorine | Protects dental enamel | Drinking water, tea, sardines | Dental caries |

disorder of zinc absorption causing failure to thrive, chronic diarrhea, alopecia, and crusted skin lesions beginning in early infancy. Zinc deficiency also develops in alcoholics because of low intake and excessive urinary excretion of zinc. Surgical patients who are receiving prolonged intravenous feeding without addition of zinc to the infusion fluid develop signs of zinc deficiency with impairment of wound healing, a generalized crusting skin eruption, and impaired sense of taste.[34-38]

### i. Copper

Copper is a component of various enzymes necessary to energy generation. It is essential to connective tissue integrity and has a role in the synthesis of the brown pigment, melanin, present in the skin and hair.

Dietary copper sources include shellfish, liver, kidney, nuts, cocoa, and fruits such as grapes and peaches. Copper deficiency may occur in infants maintained on an all-cow's milk diet and in patients on total parenteral alimentation. Deficiency leads to anemia and disorders of skin and hair pigmentation.[39,40]

A dietary requirement for other trace elements including chromium, cobalt, manganese, selenium, molybdenum, and fluorine has been identified. Functions, food sources, and signs of deficiency of these elements are summarized in Table 4.[41]

### D. Food Fortification

Food fortification has been a successful public health method of preventing man-made or environmental nutritional deficiencies. Production of iodized salt has served to prevent endemic goiter. Enrichment of bread and cereals with B vitamins was and is a major factor in the disappearance of endemic pellagra; and addition of vitamin D to milk has made rickets (vitamin D deficiency) uncommon.

Conditions which justify food fortification are (1) when through environmental factors, foods are low in trace elements essential to health and (2) when food habits or poverty prevent access to adequate nutrients by specific segments of the population.[42]

TABLE 5

**Fortified Foods**

| Food/beverage | Nutrients added |
|---|---|
| Enriched bread, pasta, and flour | Niacin<br>Thiamin<br>Riboflavin<br>Iron |
| Milk | Vitamin A<br>Vitamin D₃ |
| Margarine | Vitamin A |
| Fruit juices and drinks | Vitamin C |
| Breakfast cereals | Vitamin A<br>Vitamin D₂<br>Niacin or niacinamide<br>Riboflavin<br>Thiamin<br>Vitamin C[a]<br>Folic acid[a]<br>Vitamin B₆[a]<br>Iron |
| Salt | Iodide |

[a] For supplementation with this nutrient, see food package label.

Fortification programs may cover the entire population, but can also be designed to meet specific needs of people in particular locations, particular socioeconomic groups, or age or sex groups. Iron-fortified cereals and infant foods have been distributed to mothers and infants in the WIC program. Arguments against increasing the iron fortification of bread and cereals to control iron deficiency have been that the largest bread consumers are men who may not have need of extra iron, and some among them who may be difficult to identify may run the risk of iron overload. The potential food or beverage vehicle for fortification must be one consumed by the target population.[43] Fortified foods and the nutrients added are listed in Table 5.

### E. Recommended Dietary Allowances (RDA)

The Recommended Dietary Allowances are "the levels of intake of essential nutrients considered in the judgment of the Food and Nutrition Board on the basis of available scientific knowledge to be adequate to meet known nutritional needs of practically all healthy persons . . . RDA are recommendations for the amounts of nutrients that should be consumed daily."[44]

Recommended Dietary Allowances are shown in Table 6, which indicates the RDA by age, sex, and physiological status. The scientific bases for the RDA are limited for most water-soluble vitamins and mineral elements. Another shortcoming is that no separate RDA has yet been created for the elderly, though we cannot assume that the RDA should be the same for a person of 51 and another of 80 years. Recent workers have suggested that the inactive elderly have reduced energy requirements.

## TABLE 6

### Food and Nutrition Board, National Academy of Sciences — National Research Council Recommended Daily Dietary Allowances* (Revised 1974) (Designed for the Maintenance of Good Nutrition of Practically All Healthy People in the U.S.)

| | | Weight | | Height | | | | Fat-soluble vitamins | | | | Water-soluble vitamins | | | | | | | Minerals | | | | | |
|---|---|---|---|---|---|---|---|---|---|---|---|---|---|---|---|---|---|---|---|---|---|---|---|---|
| | | | | | | | | Vitamin A activity | | Vitamin D (IU) | Vitamin E activity (IU) | Ascorbic acid (mg) | Folacin (µg) | Niacin (mg) | Riboflavin (mg) | Thiamin (mg) | Vitamin B₆ (mg) | Vitamin B₁₂ (µg) | Calcium (mg) | Phosphorus (mg) | Iodine (µg) | Iron (mg) | Magnesium (mg) | Zinc (mg) |
| | Age (years) | (kg) | (lb) | (cm) | (in.) | Energy (kcal) | Protein (g) | (RE) | (IU) | | | | | | | | | | | | | | | |
| Infants | 0.0—0.5 | 6 | 14 | 60 | 24 | kg × 117 | kg × 2.2 | 420 | 1,400 | 400 | 4 | 35 | 50 | 5 | 0.4 | 0.3 | 0.3 | 0.3 | 360 | 240 | 35 | 10 | 60 | 3 |
| | 0.5—1.0 | 9 | 20 | 71 | 28 | kg × 108 | kg × 2.0 | 400 | 2,000 | 400 | 5 | 35 | 50 | 8 | 0.6 | 0.5 | 0.4 | 0.3 | 540 | 400 | 45 | 15 | 70 | 5 |
| Children | 1—3 | 13 | 28 | 86 | 34 | 1,300 | 23 | 400 | 2,000 | 400 | 7 | 40 | 100 | 9 | 0.8 | 0.7 | 0.6 | 1.0 | 800 | 800 | 60 | 15 | 150 | 10 |
| | 4—6 | 20 | 44 | 110 | 44 | 1,800 | 30 | 500 | 2,500 | 400 | 9 | 40 | 200 | 12 | 1.1 | 0.9 | 0.9 | 1.5 | 800 | 800 | 80 | 10 | 200 | 10 |
| | 7—10 | 30 | 66 | 135 | 54 | 2,400 | 36 | 700 | 3,300 | 400 | 10 | 40 | 300 | 16 | 1.2 | 1.2 | 1.2 | 2.0 | 800 | 800 | 110 | 10 | 250 | 10 |
| Males | 11—14 | 44 | 97 | 158 | 63 | 2,800 | 44 | 1,000 | 5,000 | 400 | 12 | 45 | 400 | 18 | 1.5 | 1.4 | 1.6 | 3.0 | 1,200 | 1,200 | 130 | 18 | 350 | 15 |
| | 13—18 | 61 | 134 | 172 | 69 | 3,000 | 54 | 1,000 | 5,000 | 400 | 15 | 45 | 400 | 20 | 1.8 | 1.5 | 2.0 | 3.0 | 1,200 | 1,200 | 150 | 18 | 400 | 15 |
| | 19—22 | 67 | 147 | 172 | 69 | 3,000 | 54 | 1,000 | 5,000 | 400 | 15 | 45 | 400 | 20 | 1.8 | 1.5 | 2.0 | 3.0 | 800 | 800 | 140 | 10 | 350 | 15 |
| | 23—50 | 70 | 154 | 172 | 69 | 2,700 | 56 | 1,000 | 5,000 | — | 15 | 45 | 400 | 18 | 1.6 | 1.4 | 2.0 | 3.0 | 800 | 800 | 130 | 10 | 350 | 15 |
| | 51+ | 70 | 154 | 172 | 69 | 2,400 | 56 | 1,000 | 5,000 | — | 15 | 45 | 400 | 16 | 1.5 | 1.2 | 2.0 | 3.0 | 800 | 800 | 110 | 10 | 350 | 15 |
| Females | 11—14 | 44 | 97 | 155 | 62 | 2,400 | 44 | 800 | 4,000 | 400 | 12 | 45 | 400 | 16 | 1.3 | 1.2 | 1.6 | 3.0 | 1,200 | 1,200 | 115 | 18 | 300 | 15 |
| | 15—18 | 54 | 119 | 162 | 65 | 2,100 | 48 | 800 | 4,000 | 400 | 12 | 45 | 400 | 14 | 1.4 | 1.1 | 2.0 | 3.0 | 1,200 | 1,200 | 115 | 18 | 300 | 15 |
| | 19—22 | 58 | 128 | 162 | 65 | 2,100 | 46 | 800 | 4,000 | 400 | 12 | 45 | 400 | 14 | 1.4 | 1.1 | 2.0 | 3.0 | 800 | 800 | 100 | 18 | 300 | 15 |
| | 23—50 | 58 | 128 | 162 | 65 | 2,000 | 46 | 800 | 4,000 | — | 12 | 45 | 400 | 13 | 1.2 | 1.0 | 2.0 | 3.0 | 800 | 800 | 100 | 18 | 300 | 15 |
| | 51+ | 58 | 128 | 162 | 65 | 1,800 | 46 | 800 | 4,000 | — | 12 | 45 | 400 | 12 | 1.1 | 1.0 | 2.0 | 3.0 | 800 | 800 | 80 | 10 | 300 | 15 |
| Pregnant | — | — | — | — | — | +300 | +30 | 1,000 | 5,000 | 400 | 15 | 60 | 800 | +2 | +0.3 | +0.3 | 2.5 | 4.0 | 1,200 | 1,200 | 125 | 18+ | 450 | 20 |
| Pregnant | — | — | — | — | — | +300 | +30 | 1,000 | 5,000 | 400 | 15 | 60 | 800 | +2 | +0.3 | +0.3 | 2.5 | 4.0 | 1,200 | 1,200 | 125 | 18+ | 450 | 20 |
| Lactating | — | — | — | — | — | +500 | +20 | 1,200 | 6,000 | 400 | 15 | 80 | 600 | +4 | +0.5 | +0.3 | 2.5 | 4.0 | 1,200 | 1,200 | 150 | 18 | 450 | 25 |

* The allowances are intended to provide for individual variations among most normal persons as they live in the U.S. under usual environmental stresses. Diets should be based on a variety of common foods in order to provide other nutrients for which human requirements have been less well defined. See text for more detailed discussion of allowances and of nutrients not tabulated.

a  Kilojoules (kJ) = 4.2 × kcal

b  Retinol equivalents.

c  Assumed to be all as retinol in milk during the first 6 months of life. All subsequent intakes are assumed to be half as retinol and half as β-carotene when calculated from International units. As retinol equivalents, three fourths are as retinol and one fourth as β-carotene.

d  Total vitamin E activity, estimated to be 80% as α-tocopherol and 20% other tocopherols. See text for variation in allowances.

e  The folacin allowances refer to dietary sources as determined by Lactobacillus casei assay. Pure forms of folacin may be effective in doses less than one fourth of the Recommended Dietary Allowance.

f  Although allowances are expressed as niacin, it is recognized that on the average 1 mg of niacin is derived from each 60 mg of dietary tryptophan.

g  This increased requirement cannot be met by ordinary diets; therefore, the use of supplemental iron is recommended.

From Recommended Dietary Allowances, 8th rev. ed., Food and Nutrition Board, National Research Council, National Academy of Science, Washington, D. C., 1974.

## F. Dietary Fiber

Current preoccupation in the U.S. in particular, and in other industrialized countries to a lesser extent, is with the fiber content of food. As stated by Macdonald, "The current widespread interest in dietary fiber is not confined to those whose main interest is in medicine and nutrition; the imagination of the lay person has been captured by the belief that fiber is a 'health food'".[45]

Dietary fiber is mainly provided by cell wall structures present in plant foods in the diet. It consists of a complex mixture of substances which have different chemical and physiological properties, sharing only the general property of being undigestible carbohydrates in human subjects. Substances included under the general heading of dietary fiber include cellulose, noncellulosic polysaccharides such as pectin or pectin substances, and lignin.[46]

Pectins are mainly supplied by fruits and vegetables such as cabbage. Cereal brans are the major sources of cellulose.

Properties of fiber components which have physiological significance can be summarized as follows: (1) high water-holding capacity, (2) cation exchange, (3) anion exchange, (4) nonionic adsorption of nutrients and foreign compounds, and (5) bacterial fermentation substrate.

Different fiber sources produce different effects when ingested because of the inherent properties of the individual fiber components, the combined effects of fiber components from any one source, the different properties of different fiber sources in the diet, and variability in the responses of individuals to dietary fiber ingestion.

From epidemiological studies, it has been claimed that high intakes of dietary fiber are protective against constipation, diverticulosis, colon cancer, and atherosclerosis.[47]

Cereal brans of large particle size increase stool weight, decrease intestinal transit time, and in colonic disease (diverticulosis) reduce colonic pressure.[48] Fine bran has a smaller effect than coarse bran in causing increases in stool weight and, in some studies, has failed to affect intestinal transit time.[49]

Studies from two areas of Denmark and Finland with a fourfold variation in colon cancer incidence have suggested that the etiology of colon cancer may be multifactorial, but that higher intakes of dietary fiber in the low cancer incidence area suggest a protective effect.[50]

Protective effects of dietary fiber against colon cancer could be due to adsorption of the carcinogen onto fiber particles with fecal excretion, or due to change in gut microflora and hence, microbial metabolism of the carcinogen. However, it is also possible that associations between high dietary fiber intake and low cancer incidence or vice versa can be explained by other factors in the diet or chemical environment which differentiate the two populations. For example, high fiber eaters tend to eat less meat and may drink less beer. Beer drinking, as well as high meat consumption, is associated with colon cancer.

Currently, two dietary fiber components are being used in prophylaxis of human health problems. Coarse bran is advocated in the prevention as well as in the treatment of constipation and is advised in diverticulosis. Bran may be given as a powder to be added to food, or the diet can be changed to contain a larger amount of bran. Pectin is hypocholesterolemic and is advised either as a semipurified product or by dietary modification in the management of patients with high serum cholesterol levels who carry a higher risk of coronary heart disease.

Dietary fiber may have positive effects other than those presently described in the standard medical or nutrition texts. Studies indicate an effect of dietary fiber, particularly coarse bran, on the upper GI tract such that gastric emptying time is slowed and adsorption of certain B vitamins including riboflavin is increased. Fiber may also promote bioavailability of certain drugs used in the treatment of acute and chronic health problems.

Adverse effects of dietary fiber on nutrient and drug bioavailability are currently being studied. It has, however, already been shown that high intake of cereal bran may depress iron absorption. The suggestion has also been made by Mendeloff[51] that there may be an optimal dose for dietary fiber, and that if this dose is exceeded, unwanted side effects may result.

### G. Food Additives

Food additives are integral components of our diet. Additives consist in nutrients and nonnutrient chemicals. Functions of food additives are as follows:

1.  To improve nutritive value of foods (food fortification)
2.  To enhance flavor
3.  To maintain appearance
4.  To preserve food
5.  To impart desired consistency
6.  To control acidity or alkalinity
7.  To give the desired color
8.  To act as maturing and bleaching agents
9.  For artificial sweetening
10. To perform special functions, e.g., prevention of caking or moisture retention

Types of food additives are as follows:

1.  Nutritional supplements, e.g., vitamins and minerals in cereals and breads, iodide or iodate salt and bread, fluoride in drinking water
2.  Flavoring agents, e.g., salt, sugar, volatile oils, spices, monosodium glutamate, amyl acetate
3.  Preservatives, e.g., antibiotics, antioxidants, and derivatives of proprionic acid
4.  Consistency improving agents, e.g., emulsifiers such as lecithin or methylcellulose
5.  Coloring agents, e.g., natural and synthetic dyes
6.  Bleaching agents, e.g., chlorine or potassium bromate
7.  Reaction controlling agents, e.g., sodium bicarbonate, citric and tartaric acids
8.  Humectants (agents to improve water-holding capacity), e.g., glycerine
9.  Anticaking agents, e.g., magnesium carbonate
10. Curing agents, e.g., nitrates and nitrites
11  Sweetening agents, e.g., saccharine

Food additives are defined, then, as any substance which may become a part of food. Advantages of food additives are in the improvement of the nutritional value of foods and food preservation. Irrational concerns about chemical food additives have been pointed out by Jukes.[52] Concerns among organic and health food movements have developed the idea that all food additives are bad. These groups have also confused the public by suggesting that synthetic vitamins are less valuable nutritionally than their natural counterparts. In fact, synthetic vitamins are identical with naturally occurring vitamins. Avoidance of food additives as a means of controlling hyperactivity in children has been advocated but is not proven to be beneficial. The most injurious of all food additives is additional food eaten beyond energy needs.

Risk benefit relationships for food additives should be considered under the following headings:

1. Estimated hazard to the consumer
2. Consumer needs and wishes
3. Requirement of the food supply for public health
4. Needs of the food producer and processer
5. Economic considerations
6. Availability of methods and mechanisms for regulatory control

UN/FAO/WHO recommendations for a food additive are that:

1. It should be technologically effective.
2. It should be safe in use.
3. It should not be used in excess of the amount required to achieve the desired effect.
4. It should not be used to deceive the customer.
5. It should not be used to decrease the nutritive value of the food.

Special health risks from food additives which have been identified are as follows:

1. Allergic reactions, e.g., to the yellow food-coloring agent tartrazine
2. Induction of resistant microorganisms, i.e., to antibiotics
3. Carcinogenicity, e.g., from the formation of nitrosamines or from use of diethyl-stilbesterol (DES) in foods
4. Mutagenicity or teratogenicity (production of fetal malformations), e.g., by DES or by food-coloring agents
5. Malabsorption, e.g., by use of modified starches in infant foods
6. Acute toxic reactions, e.g., from ingestion of nitrite or monosodium glutamate

Acute toxic reactions to food additives show extreme interindividual variability. In some individuals, intake of nitrite or nitrate in foods containing these additives, e.g., hot dogs, cold cuts, and ham, causes headaches due presumably to dilatation of the blood vessels around the brain. The Chinese restaurant syndrome (CRS) occurs in certain people when they eat foods containing monosodium glutamate. Symptoms are headaches, tightness in the chest, and a burning sensation in the back of the neck. Sometimes these symptoms are associated with a feeling of weakness. The effects are transient. It seems that men are more likely to be affected than women. Large intakes of monosodium glutamate are likely if oriental foods (Japanese or Chinese) are eaten.

International recommendations have been made for the limitaton of food additives for infants' foods. The risks of food additives for infants are that (1) metabolism of the additives to produce nontoxic end products may be inadequate, (2) the infant intake of food additives may be large relative to their body weight, and (3) infants run the risk of development of allergy to the additives.

Recommendations are that (1) no foods containing additives should be given to infants under the age of 3 months (the author recommends that food additives not be given to infants under 6 months) (2) foods containing nitrates or nitrites should be avoided in this age group, and (3) there should be limited use of modified starches such as food thickeners.

Age has a marked effect on the potential acute toxicity of inorganic nitrates and nitrites, such that infants are susceptible to the acute toxic effects of these additives. Infants under 6 months who ingest foods containing nitrite or nitrate are at risk for the development of methemoglobinemia. Methemoglobinemia is a condition in which the hemoglobin in the blood is oxidized. In the young infant, ingested inorganic nitrate is converted to nitrite by bacterial action in the gastrointestinal tract. Then the endogenously formed nitrite or ingested nitrite is absorbed into the circulation and the immature (fetal) hemoglobin is oxidized to methemoglobin which may cause cyanosis. Reduction of the methemoglobin so formed to normal hemoglobin is slow in the infant.[53,54]

## H. Food Intake and Dietary Requirements in Pregnancy and Lactation

Energy and nutrient requirements are increased in the pregnant woman, both to insure optimum development of her body for childbearing and to supply the needs of the developing fetus. Food aversions and cravings, which are traditionally associated with pregnancy, are not reliable indicators of a nutritionally adequate diet. While it has been shown that women may decrease consumption of alcoholic beverages and sodas in the first 3 months of pregnancy because of concern for the baby's development, a desire for particular foods and beverages appears to be influenced more by cultural, geographical, and personal factors. In an Albany study by Hook, common cravings cited by pregnant women were for ice cream, chocolates, other candy, and various fruits, whereas the most frequent aversions were for meats and poultry.[55] Approximately 50% of the women interviewed in this study indicated that they increased their consumption of milk during the first half of pregnancy because of concern for their own health or for the baby's health. Other women explained their higher milk consumption as being due to a craving or due to the better taste of milk in pregnancy. Although it has been suggested that the increased desire for candy and ice cream in pregnancy could be to supply energy and calcium needs, this notion is not well grounded in fact.

An earlier study by Edwards and co-workers showed that among black women in Tuskegee, Alabama, common cravings were for clay, fish, chicken, cornstarch, and apples, rather than for candy or other fruits.[56] These food desires, while traditional to southern black women, were not commonly cited by black women in the Albany sample. We are impressed that these studies support the view that local custom defines food desires in pregnancy. While change in the sensory appeal of food may influence cravings or aversions, this is currently an untested theory.

Eating clay or laundry starch in pregnancy or at other times may impair iron absorption and must be discouraged.[57]

The actual energy requirement of a pregnant woman can only be calculated by reference to her physical activity and perhaps, in the adolescent girl, to nonpregnant growth requirements as well as to the total energy cost of pregnancy. An addition of 300 kcal per day has been estimated to represent additional basal energy needs. An easy approach to the determination of whether a pregnant woman is getting enough food for her needs is to examine changes in body weight. Total weight gain during pregnancy should be about 20 to 22 lb. This gain should follow a pattern such that in the first third (first trimester) of pregnancy, weight gain is 2 to 5 lb, and thereafter weight gain is between 2½ and 3 lb a month.

Definitions of underweight deviations and guidelines for weight gain in pregnancy have been laid down by Pitkin as follows:[58]

- Underweight: prepregnancy weight 10% or more below standard weight for height and age
- Overweight: prepregnancy weight 20% or more above standard weight for height and age
- Inadequate pregnancy gain: less than 1 kg (2.2 lb) per month during the second and third trimesters of pregnancy

• Excessive pregnancy gain: gain of more than 3 kg (6.6 lb) per month during this period

Underweight patients are more likely to have low birth weight infants, and obese patients are more at risk for hypertension and diabetes.

Dietary restriction during pregnancy is usually inadvisable and should certainly be discouraged when there is concern that by so doing the woman will decrease the nutritional quality of her diet. There are increased requirements for nutrients during pregnancy. Particular daily needs are for additional protein, iron, folacin (folic acid), calcium, and vitamin D. Intake of other nutrients should be modestly increased. Dietary guidelines for pregnant women are shown in Table 7.

Iron deficiency anemia is common during pregnancy. In order to avoid the risk of iron deficiency anemia, a daily intake of iron of 30 to 60 mg per day is recommended. In order to achieve this intake, iron supplements must be given. Since folacin requirements during pregnancy are also high due to fetal developmental needs, the daily nutrient supplement should contain both iron and folic acid, the latter at levels of between 400 and 800 μg per day.[59]

Risk factors for inadequate nutrition are most frequently present in adolescent girls.[60-63] Inadequate nutrition in these girls is determined by:

1. Lack of nutrition knowledge
2. Rejection by the parents
3. Desire to hide the pregnancy
4. Growth requirements
5. Unwanted pregnancy
6. Inadequate antenatal care
7. Lack of funds
8. Physician or other health professional recommends weight control

Antenatal care must include nutritional screening, nutrition education, and nutritional counseling. It is further of the utmost importance that nutrition education and counseling should stress not only a desirable eating pattern to achieve optimum nutrition for the patient and her baby, but also that food or substance abuse must be avoided. Under food abuse we include excessive intake of salt, sugar, sodas, coffee, candy, and other foods which either pose a long-term health risk or detract from the intake of other foods of higher nutrient value. Further, girls and women should be counseled to refrain from eating nonfood substances such as clay or laundry starch, both because of adverse effects on nutrient absorption and also because large intakes of these substances may reduce dietary quality.

The same dietary principles pertain during lactation as during pregnancy. Recommended intakes of nutrients are given in Table 8.

TABLE 7

Dietary Guidelines for Pregnant Women (See Table 6)

1. Add 300 kcal per day to prenatal diet, meeting RDA for age.
2. Drink 1 qt of milk or eat the equivalent in cheese to meet RDA for calcium and additional protein needs.
3. Increase protein intake by 30 g per day using protein of high biological value.
4. Eat four to five servings of fruit plus vegetables per day, including two citrus, one green and one yellow vegetable (citrus fruit can be replaced by citrus fruit juice).
5. If dietary record shows iron and/or folacin intake less than RDA for pregnant women, give physiological supplement.

TABLE 8

Recommended Energy, Food, and Nutrient Intakes for Lactating Women (See Table 6)

1. Add 500 kcal per day to nonlactating energy needs.
2. Increase protein intake by 20 g per day using protein sources of high biological value.
3. Drink 1 qt of milk per day.
4. Increase servings of fruit (or citrus fruit juices) and vegetables to six per day to supply RDA for ascorbic acid, folacin, and part of the RDA for vitamin A.

Milk intake as a calcium, vitamin D, and riboflavin source should be at a level of 1 to 2 pt per day. Alcoholic beverages should not be taken, or only taken occasionally. Drugs should not be taken except when essential to the treatment of disease.

Physiological weight loss occurs after childbirth. If a woman has gained excessive weight during pregnancy, total caloric intake should not exceed requirements. However, it is injudicious to prescribe a weight-reducing diet during lactation, because the nutrient quality of the diet may be insufficient for the mother's and infant's needs.[64]

## I. Human Milk and Breast-Feeding

Under the definition of human milk we include colostrum, transitional human milk, and mature human milk. Colostrum is the breast secretion which is available to the newborn infant in the first 3 days after birth. Amounts are small but the value to the infant from colostrum is mainly due to its rich content of bacterial and viral antibodies which confer protection to the infant against infections by pathogenic organisms which enter the body via the gastrointestinal tract.[65]

Transitional human milk is the milk secreted between the 3rd and 10th day after childbirth. Variability exists in the composition of mature human milk, the approximate composition of representative samples being shown in Table 9.[66] The macronutrient (protein, fat, carbohydrate) composition of human milk is not greatly affected by the maternal diet, but the vitamin content of the milk reflects maternal intake. Poorly nourished women secrete less milk.[67,68]

Mature human milk is an almost ideal food for the infant. However, vitamin D content is low, and neither iron nor fluoride are provided. Breast-fed infants therefore require supplements of these nutrients. Advantages of breast-feeding are[69]

1. Economic: the cost of formula feeding may persuade very low-income women to feed overdilute formulas.
2. Breast-feeding reduces the risk of intestinal infections which may be due to contamination of the formula.
3. Breast-feeding, particularly allowing the infant to obtain colostrum, provides protection against enteric and other infections.
4. A higher level of contact between the mother and the infant may have beneficial effects on behavioral development.
5. Breast-feeding may decrease the risk of obesity because the volume and caloric density of feeds cannot be forced.

Contraindications to breast-feeding are when the mother has to take medications which would be toxic to the infant.[70] Other contraindications are when the mother has cancer of the breast or breast infection. Low birth-weight infants in perinatal units should be given expressed breast milk.

In 1976 the Committee on Nutrition of the American Academy of Pediatrics (AAP) made the following recommendations to encourage breast-feeding:[71]

TABLE 9

Composition of Mature Human Milk

| Composition | Human milk | Composition | Human milk |
|---|---|---|---|
| Water (ml/100 ml) | 87.1 | Copper (μg) | 400 |
| Energy (kcal/100 ml) | 75 | Zinc (mg) | 3—5 |
| Total solids (g/100 ml) | 12.9 | Iodine (μg) | 30 |
| Protein (g/100 ml) | 1.1 | Selenium (μg) | 13—50 |
| Fat (g/100 ml) | 4.5 | Iron (mg) | 0.5 |
| Lactose (g/100 ml) | 6.8 | Vitamins per liter | |
| Ash (g/100 ml) | 0.2 | Vitamin A (IU) | 1898 |
| Major minerals per liter | | Thiamin (μg) | 160 |
| Calcium (mg) | 340 | Riboflavin (μg) | 360 |
| Phosphorus (mg) | 140 | Niacin (μg) | 1470 |
| Sodium (mEq) | 7 | Pyridoxine (μg) | 100 |
| Potassium (mEq) | 13 | Pantothenate (mg) | 1.84 |
| Chloride (mEq) | 11 | Folacin (μg) | 52 |
| Magnesium (mg) | 40 | $B_{12}$ (μg) | 0.3 |
| Sulfur (mg) | 140 | Vitamin C (mg) | 43 |
| Trace minerals per liter | | Vitamin D (IU) | 22 |
| Chromium (μg) | — | Vitamin E (mg) | 1.8 |
| Manganese (μg) | 7—15 | Vitamin K (μg) | 15 |

Data on proximate composition are from: Macy, I. G. and Kelly, H. J., Human milk and cow's milk in infant nutrition, in *Milk: The Mammary Gland and Its Secretion*, Vol. 2, Kon, S. K. and Cowie, A. T., Eds., Academic Press, New York, 1961, 265.

From: Fomon, S. J. and Filer, L. J., Jr., Milks and formulas, in *Infant Nutrition*, 2nd ed., Fomon, S. J., Ed., W. B. Saunders, Philadelphia, 1974, 362. (With permission.)

1.  Provision of educational programs for adolescents and pregnant women
2.  Reinforcement of advantages of breast-feeding by obstetricians, pediatricians, and nurses attending pregnant women
3.  Changes in employment policies and working conditions to make breast-feeding practical for working mothers (it was recommended that day-care centers be provided at or near places of work)

The Committee supported a policy to prohibit advertising or promotional practices which might discourage mothers from breast-feeding.

The Committee proposed a single standard for nutrient composition of milk-based and milk-substitute formulas. Finally, the Committee recommended that the FDA create two categories of infant formulas: one for normal infants for which the standards are proposed, and one for infants with special medical conditions.

## J. Formula Diets and Dietary Supplements

### 1. Infant Formulas

In the U.S. it is accepted that formulas for normal infants provide 7 to 16% of food energy from protein, 30 to 55% of food energy from fat, and the remainder of food energy from carbohydrate sources.[72] Linoleic acid, a source of essential fatty acid, should be provided at a level to supply 1% of food energy in the formula.[73] The minimum content of vitamins and minerals recommended by the Committee on Nutrition of the American Academy of Pediatrics and accepted by the U.S. Food and Drug Administration is indicated in Table 10.[74,75]

TABLE 10

**Nutrient Levels of Infant Formulas (per 100 kcal)**

| Nutrient | FDA 1971 Regulations (minimum) | CON 1976 Recommendations: Minimum | Maximum |
|---|---|---|---|
| Protein (g) | 1.8 | 1.8 | 4.5 |
| Fat | | | |
| (g) | 1.7 | 3.3 | 6.0 |
| (% cal) | 15.0 | 30.0 | 54.0 |
| Essential fatty acids (linoleate) | | | |
| (% cal) | 2.0 | 3.0 | — |
| (mg) | 222.0 | 300.0 | — |
| Vitamins | | | |
| A (IU) | 250.0 | 250.0 (75 µg)[a] | 750.0 (225 µg)[a] |
| D (IU) | 40.0 | 40.0 | 100.0 |
| K (µg) | — | 4.0 | — |
| E (IU) | 0.3 | 0.3 (With 0.7 IU/g linoleic acid) | — |
| C (Ascorbic acid) (mg) | 7.8 | 8.0 | — |
| B₁ (Thiamin) (µg) | 25.0 | 40.0 | — |
| B₂ (Riboflavin) (µg) | 60.0 | 60.0 | — |
| B₆ (Pyridoxine) (µg) | 35.0 | 35.0 (With 15 µg/g of protein in formula) | — |
| B₁₂ (µg) | 0.15 | 0.15 | — |
| Niacin | | | |
| (µg) | — | 250.0 | — |
| (µg equivalent) | 800.0 | — | — |
| Folic acid (µg) | 4.0 | 4.0 | — |
| Pantothenic acid (µg) | 300.0 | 300.0 | — |
| Biotin (µg) | — | 1.5 | — |
| Choline (mg) | — | 7.0 | — |
| Inositol (mg) | — | 4.0 | — |
| Minerals | | | |
| Calcium (mg) | 50.0[b] | 50.0[b] | — |
| Phosphorus (mg) | 25.0[b] | 25.0[b] | — |
| Magnesium (mg) | 6.0 | 6.0 | — |
| Iron (mg) | 1.0 | 0.15 | — |
| Iodine (µg) | 5.0 | 5.0 | — |
| Zinc (mg) | — | 0.5 | — |
| Copper (µg) | 60.0 | 60.0 | — |
| Manganese (µg) | — | 5.0 | — |
| Sodium (mg) | — | 20.0 (6 mEq)[c] | 60.0 (17 mEq)[c] |
| Potassium (mg) | — | 80.0 (14 mEq)[c] | 200.0 (34 mEq)[c] |
| Chloride (mg) | — | 55.0 (11 mEq)[c] | 150.0 (29 mEq)[c] |

[a]   Retinol equivalents.
[b]   Calcium to phosphorus ratio must be no less than 1.1 nor more than 2.0.
[c]   Milliequivalent for 670 kcal/*l* of formula.

From Special report: commentary on breast-feeding and infant formulas, including proposed standards for formulas, *Nutr. Rev.*, 34, 1976, 251. With permission.

The composition of commercially available milk-based formulas is shown in Table 11. These formulas are generally available as powdered products, as concentrates (133 kcal/100 m*l*), or as ready-to-feed liquids (67 kcal/100 m*l*). Products so indicated are available with iron fortification at a level of 12 to 13 mg iron per liter. It has been

## TABLE 11

### Composition of Milk-Based Formulas

|  | Infants | Older infants |
|---|---|---|
| Components | Nonfat cow's milk | Nonfat milk solids |
| Protein | Vegetable oil | Corn oil |
| Fat | Lactose or corn | Corn syrup |
| Added carbohydrate | syrup solids | |

|  | A | B | C |
|---|---|---|---|
| Macronutrients (per *l*) | 13.7 | 11.8 | 28 |
| Protein, g | 35.5 | 27.4 | 20 |
| Fat, g | 49.2 | 55.7 | 62 |
| Carbohydrate, g | | | |
|  | (Ready to feed at 20 kcal per fluid oz) | | (Ready to feed at 16 kcal per fluid oz) |

| Vitamins per *l* | | |
|---|---|---|
| Vitamin A, IU | 1690—2500 | 2400 |
| Vitamin D, IU | 400—400 | 400 |
| Vitamin E, IU | 12.7—15 | 12 |
| Vitamin C, mg | 55—55 | 50 |
| Folic acid (folacin) $\mu$g | 50—100 | 100 |
| Thiamin, $\mu$g | 529—650 | 750 |
| Riboflavin, $\mu$g | 634—1000 | 900 |
| Niacin equivalent, mg | 7—8.5 | 10 |
| Vitamin B$_6$, $\mu$g | 400—423 | 600 |
| Vitamin B$_{12}$, $\mu$g | 1.5—2 | 2.5 |
| Pantothenic acid, mg | 3—3 | 4 |
| Vitamin K$_1$, $\mu$g | —[a] 27 | —[a] |
| **Minerals per *l*** | | |
| Calcium, g | 0.51—0.55 | 0.80 |
| Phosphorus, g | 0.39—0.46 | 0.60 |
| Magnesium, mg | 41 | 80 |
| Zinc, mg | 4—5 | 6 |
| Copper, mg | 0.6—0.41 | 0.9 |
| Iodine, $\mu$g | 69—100 | 60 |
| Manganese, $\mu$g | 1—34 | —[a] |
| Iron, mg | 1.5—Tr | 18 |
| Sodium, g | —[a] —[a] | 0.45 |

*Note:* A = Enfamil, Mead-Johnson Nutrition Division.
B = Similac, Ross.
C = Similac Advance, Ross.

[a] Value not given.

recommended by the AAP of 1971 that formula-fed infants should receive iron-fortified formulas to prevent iron deficiency.

Formulas with added whey protein, modified to approximate the composition of human milk, are shown in Table 12. Fomon and Filer have recommended the use of formulas containing partially demineralized whey protein for infants with renal disease or congenital heart disease, or in any other condition in which it is desirable to supply high-energy feedings without an excess of sodium potassium or chloride.[68]

The composition of soy-based formulas (with protein from water-soluble soy isolates), suitable for infants with milk protein allergy, are shown in Table 12. Special

TABLE 12

**Composition of Infant Formulas**

| Components | Whey based | Soy based |
|---|---|---|
| Protein | Lactalbumin (whey protein demineralized) | Soy isolate |
| Fat | Safflower oil | Vegetable oil |
| Carbohydrate | Lactose | Corn syrup solids or sucrose |
| | **A** (w/v) | **B** |
| Protein | 1.5% | 20 g |
| Fat | 3.6% | 36 g |
| Carbohydrate | 7.2% | 68 g |
| | (Ready to feed, 20 kcal per fluid oz) | (Ready to feed, 20 kcal per fluid oz) |
| Vitamins per *l* | | |
| Vitamin A, IU | 2536 | 2500 |
| Vitamin D, IU | 423 | 400 |
| Vitamin E, IU | 9.5 | 15 |
| Vitamin C, mg | 58 | 55 |
| Folacin, $\mu$g | 53 | 100 |
| Thiamin, $\mu$g | 700 | 400 |
| Riboflavin, $\mu$g | 1057 | 600 |
| Niacin equivalent, mg | 10.0 | 9.0 |
| Vitamin $B_6$, $\mu$g | 422 | 400 |
| Vitamin $B_{12}$, $\mu$g | 1 | 3 |
| Pantothenic acid, mg | 2 | 5 |
| Vitamin $K_1$, $\mu$g | 58 | 150 |
| Minerals per *l* | | |
| Calcium, g | 0.4 | 0.7 |
| Phosphorus, g | 0.33 | 0.5 |
| Magnesium, mg | 53 | 50 |
| Zinc, mg | 4 | 5 |
| Copper, mg | 0.48 | 0.5 |
| Iodine, $\mu$g | 68 | 150 |
| Manganese, $\mu$g | 158 | 200 |
| Iron, mg | 12.7 | 12 |
| Sodium chloride, g | 0.15 | 0.3 |

*Note:*  A  = SMA, Wyeth.
        B  = Similac Isomil, Ross.

formulas are available for infants with metabolic disorders and malabsorption syndromes. Nutramigen ® (Mead Johnson Nutritional Division) consists of a casein hydrolysate with added sugar, starch, corn oil, minerals, and vitamins. This formula is used for feeding infants or young children with galactosemia, as well as infants with milk protein allergy.[76] An alternate formula for galactosemic infants is Meat Base Formula (MBF, Gerber ®), which is prepared from beef heart with added fat and carbohydrate.

Lofenalac (Mead Johnson Nutritional Division), produced from a casein hydrolysate with reduction in phenylalanine content, is the formula of choice for infants and children with phenylketonuria. In order to meet the growth requirements of young phenylketonuric patients, Lofenalac and another source of low phenylalanine protein should be given.

Portagen® and Pregestimil® (Mead Johnson Nutrition Division) which are formulas containing medium chain triglycerides can be used for infant feeding. Indications for use of Portagen® are in fat maldigestion and malabsorption syndromes in which there is a failure to digest or absorb dietary fat.[77] Pregestimil® is indicated in the nutritional management of infants who have had intestinal resections, in infants with cystic fibrosis, in chronic diarrheas, and in various severe malabsorption syndromes. The composition of these special formulas, uses, precautions in administration, and mode of preparation are described in the *Physicians' Desk Reference.*[78]

### 2. Total Feeding Formulas — Oral or Tube Feeding
#### a. Chemically Defined (Elemental Diets)
Chemically defined diets may be based on (1) nutrients requiring little or no digestion, (2) nutrients which are preferentially absorbed, and (3) formulas in which specific macronutrients or micronutrients are omitted.

Examples of chemically defined diets for total meal replacement or diet supplement are shown in Table 13. Chemically defined diets may be given by mouth or by feeding tube. Indications are as follows: (1) maldigestion or malabsorption syndromes with temporary or permanent intolerance of dietary fat, protein, or lactose, (2) correction of a catabolic state, particularly prior to surgery, (3) late stage renal disease, and (4) acute or chronic liver failure. Specific advantages of chemically defined diets are that utilization of the administered nutrients can occur when digestive and absorptive function are reduced. Further, the ultra-low residue characteristic of these formula diets offers certain advantages when preparation for bowel surgery is being undertaken.

Shortcomings and side effects of chemically defined diets are (1) lack of palatability, (2) gastric retention, (3) diarrhea, (4) constipation, and (5) nutritional deficiency (micronutrients normally present in the diet may be absent from artificially prepared formulas). Particular risks are for trace element deficiencies.

#### b. Tube Feedings
Tube feeding is recommended for patients with conditions which make oral intake of food difficult or impossible, but in whom the digestive and absorptive functions of the gastrointestinal tract are maintained or partially maintained. Patients requiring tube feeding are (1) those with neurological diseases which prevent normal swallowing without food regurgitation or aspiration, (2) those with cancer or severe inflammatory disease of the mouth, throat, tongue, or neck where oral ingestion is precluded by pain, and (3) those with acute or late effects of injury or surgery to the jaws, mouth, pharynx, and larynx, esophagus, or stomach, which cause obstruction or discontinuity of the upper gastrointestinal tract.

Tube feedings are sometimes employed as a sole method of feeding or as an adjunct method of feeding in patients who refuse to eat because of depression, anorexia nervosa, or lack of desire for survival. Justification for tube feeding in these conditions is limited because the procedure may be interpreted by the patient as a punitive activity. Tube feeding should never be undertaken except under very close medical supervision in unconscious or semiconscious patients, because of the risk that the tube may pass into the upper respiratory tract or that food may be regurgitated and inhaled, leading to aspiration pneumonia.

Tube feedings may be given via the mouth or via a gastrostomy or jejunogastrostomy tube. Tube feedings may be intermittent or continuous. If the latter technique is adopted, a pumping device is recommended for long-term maintenance.

Tube feedings may be prepared from (1) blenderized conventional foods prepared in the hospital or home, (2) commercially available complete meal replacements based

TABLE 13

Examples of Chemically Defined Diets

| | Precision isotonic diet per 1500 kcal (Doyle Pharmaceuticals) | Flexical diet per 2000 kcal (Mead-Johnson) |
|---|---|---|
| Protein, g | 45[a] | 45[b] |
| Carbohydrate, g | 225[c] | 305[d] |
| Fat, g | 47[e] | 67.9[f] |
| Vitamin A, IU | 5000 | 5000 |
| Vitamin D, IU | 400 | 400 |
| Vitamin E, IU | 30 | 45 |
| Vitamin C, mg | 90 | 300 |
| Folic acid, mg | 0.4 | 0.4 |
| Thiamin, mg | 2.25 | 3.8 |
| Riboflavin, mg | 2.6 | 4.3 |
| Niacin, mg | 20 | 50 |
| Vitamin $B_6$, mg | 3.0 | 5.0 |
| Vitamin $B_{12}$, μg | 6.0 | 15 |
| Biotin, mg | 0.3 | 0.3 |
| Pantothenic acid, mg | 10 | 25 |
| Calcium, g | 1.0 | 1.2 |
| Phosphorus, g | 1.0 | 1.0 |
| Iodine, μg | 150 | 150 |
| Iron, mg | 18 | 18 |
| Magnesium, mg | 400 | 400 |
| Copper, mg | 2.0 | 2.0 |
| Zinc, mg | 15 | 20 |
| Choline, mg | 100 | 500 |
| Potassium, g | 1.5 | 2.5 |
| Sodium, g | 1.2 | 0.7 |
| Chloride, g | 1.6 | 2 |
| Manganese, mg | 4.0 | 5 |
| Vitamin K, μg | 100 | 250 |

[a]   Egg white solids.
[b]   Enzymatically hydrolyzed casein + L-methionine, L-tyrosine, and L-tryptophan.
[c]   Sucrose.
[d]   Medium-chain triglycerides (fractionated coconut oil).
[e]   Vegetable oil.
[f]   Corn syrup solids.

on milk or meat, vegetable oils, and cereals, and (3) chemically defined diets.

Risks of tube feeding are (1) injury to the pharynx or esophagus, (2) gastric dilatation, (3) dumping syndrome, (4) aspiration pneumonia, (5) asphyxia, and (6) nutritional deficiencies (see chemically defined diets).

General nutritional principles which must be considered in oral or tube feeding of formulas are as follows:

1.  The total caloric value of the formula must meet the patient's needs.
2.  If the formula contains a protein as the nitrogen source, then the protein should be well tolerated and of high biological value.
3.  If the defined chemical formula contains an amino acid mixture, then this mixture must be such as to produce and maintain nitrogen balance.

4. Carbohydrates present in the formula must be tolerated by the patient. If the patient is lactose intolerant, or intolerant of sucrose or maltose, these disaccharide sugars must not be present in this formula.
5. The vitamin content of the formula must be known. Either sufficient vitamins must be provided by the formula to maintain or restore vitamin-dependent function, or supplementary vitamins must be given.
6. Deficient minerals must be supplied for the patient's needs but electrolyte or other mineral imbalance or overload must be avoided.
7. Sufficient fluid must be supplied to provide for excretion of nitrogenous end products and other waste materials excreted by the kidney.

#### c. Parenteral Hyperalimentation

Indications for intravenous hyperalimentation are as follows:

1. Major abnormalities of the gastrointestinal tract
   a. When ingestion of food is impossible
   b. When there is an inability to retain ingested food
   c. When there is severe malabsorption or maldigestion
   d. With trauma to the gastrointestinal tract
   e. With elective surgery on the gastrointestinal tract
2. Preparation for surgery in nutritionally depleted patients
3. After surgery or injury including burns associated with severe catabolic losses
4. As supplementation to oral feeding in cancer patients
5. In renal or hepatic failure

In infants, special indications for parenteral hyperalimentation are

1. Surgical disease of the gastrointestinal tract secondary to major malformations
2. Chronic intractable diarrhea
3. Very low birth weight

Sources of nutrients incorporated into intravenous solutions for hyperalimentation are
1. Food energy sources: glucose, fructose, lipids, or alcohol
2. Nitrogen source: crystalline amino acid mixtures or protein hydrolysates
3. Electrolytes: sodium, potassium, and chloride
4. Minerals: calcium, magnesium, and phosphorus
5. Trace minerals: zinc, copper, etc.
6. Fat and water-soluble vitamins
7. Essential fatty acids

The composition of infusates for adult and pediatric use are shown in Tables 14 and 15. Routes of administration are central via a central venous catheter and via a peripheral venous infusion.

For techniques, clinical application, and complications of parenteral hyperalimentation, readers should refer to specialized texts.[79-84]

#### K. Diseases Which Alter Nutrient Requirements

Diseases which alter nutrient requirements may be inborn (congenital) or acquired.

## TABLE 14

### Solution for Parenteral Nutrition

| Component | Amount/$l$ | Amount/day[a] |
|---|---|---|
| Glucose (monohydrate) | 250 g | 750 g |
| Amino acids | 28.2 g | 8.5 g |
| Nitrogen | 24.7 g | 12.5 g |
| Glucose | 900 cal | 2700 cal |
| Na+ | 33 mEq[b] | 99 mEq |
| K+ | 40 mEq | 120 mEq |
| Mg++ | 10 mEq | 30 mEq |
| Ca++ | 4.5 mEq | 13.5 mEq |
| CL | 30 mEq | 90 mEq |
| Acetate | 25 mEq | 75 mEq |
| P (inorganic phosphorus) | 420 mg | 1260 mg |
| ZnSO4 | 5 mg | 15.0 mg |
| Vitamins[c] | 1.4 m$l$ | 4.2 m$l$ |

[a]   Amount in 3 $l$ given to an adult patient daily.

[b]   30 mEq/$l$ are added as sodium chloride and 3 mEq/$l$ are in FreAmine as sodium bisulfate.

[c]   Vitamins are given as aqueous multivitamin infusion solution (MVI 10 m$l$-ampule). Folate and B$_{12}$ are added to the infusate separately. Iron as Imferon is given separately when needed.

From Meng, H. C., Parenteral nutrition: principles, nutrient requirements, techniques, and clinical applications, in *Nutritional Support of Medical Practice*, Schneider, H. A., Anderson, C. E., and Coursin, D. B., Eds., Harper & Row, Hagerstown, Maryland, 1977, 163.

## TABLE 15

### Usual Composition of Infusate

| Component | Daily intake |
|---|---|
| Nitrogen source | 2.5 g/kg |
| Glucose | 25—30 g/kg |
| NaCl | 3—4 mM/kg |
| KH$_2$PO$_4$ | 2—3 mM/kg |
| Ca gluconate | 0.25 mM/kg (0.5 mEq/kg) |
| MgSO$_4$ | 0.125 mM/kg (0.25 mEq/kg) |
| MVI[a] | 1 m$l$ |
| Vitamin B$_{12}$ | 10 µg |
| Folic acid | 50—75 µg |
| Vitamin K$_1$ | 250—500 µg |
| Total volume | 130 m$l$/kg |

[a]   U.S. Vitamin Corp., Tuckahoe, N.Y.

From Heird, W. C. and Winters, R. W., Parenteral nutrition: Pediatrics, in *Nutritional Support of Medical Practice*, Schneider, H. A., Anderson, C. E., and Coursin, D. B., Eds., Harper & Row, Hagerstown, Maryland, 1977, 185.

## 1. Inborn Errors of Metabolism

In infants inborn errors of metabolism such as phenylketonuria (PKU) and galacto-semia are enzyme deficiencies which preclude normal utilization of specific nutrients in the body. These nutrients then accumulate and are toxic. In PKU, the deficiency is of the enzyme phenylalanine hydroxylase. Infants born with this disease cannot toler-ate phenylalanine. If they receive phenylalanine greater than the minimal level to allow protein synthesis, mental deficiency and severe behavioral abnormalities develop. Since phenylalanine is present in all protein foods, these foods have to be restricted and a low-phenylalanine formula is given to supply other amino acids. If the diet of these infants is too low in phenylalanine, growth failure occurs, as well as skin change, due to inadequate protein synthesis. Tolerance for phenylalanine increases with age in chil-dren with PKU. However, in adult women with PKU, counseling is needed to insure that their phenylalanine intake is restricted during pregnancy. Pregnant women with PKU, who fail to follow their prescribed low-phenylalanine diets, develop high blood levels of phenylalanine which is the cause of mental retardation in their offspring.[85,86]

Galactosemia is a congenital disease in which the sugar galactose cannot be metab-olized. Galactosemic infants are intolerant of lactose which contains galactose. There-fore, they cannot be given milk, milk products, or any other foods containing lactose or galactose. If fed milk as infants, they vomit, fail to thrive, and develop jaundice, cataracts, and mental retardation. Prevention of growth retardation as well as liver, eye, and mental manifestations of the disease requires early detection as well as vigi-lance in respect to adherence to the galactose-lactose-free diet.[87]

Acrodermatitis enteropathica, a disease which is characterized by growth retarda-tion, diarrhea, hair loss, and skin changes, is due to an inborn defect of zinc absorp-tion. It is controlled by zinc supplements.[88]

## 2. Acquired Diseases

Acquired diseases which increase food energy and nutrient requirements, are acute and chronic infections, diseases of the gut or pancreas associated with maldigestion or malabsorption, protein energy malnutrition, avitaminoses, iron deficiency anemia, folic acid deficiency, pernicious anemia, hemolytic anemia (e.g., Sickle cell anemia), alcohol-related diseases, epilepsy, and cancer. Common diseases which decrease or alter nutrient requirements are coronary heart disease, hypertension, chronic renal dis-ease, and congestive heart failure.[89,90] Common diseases and disease categories which increase or decrease nutrient requirements are listed in Table 16.

## L. Nutritional Assessment

Whereas all health professionals have need to make nutritional assessments and to determine the nutrition or food-related requirements of their patients or clients, objec-tives and means of the assessments differ. Goals of nutritional assessment are diagnosis by physicians; determination of precise nutritional status by physicians and nurse cli-nicians; determination of nutritional risk by physicians, nurses, pharmacists, dentists, and medical social workers; diet prescription by physicians, nurse clinicians, and die-titians; nutrition education by nutritionists, health educators, physicians, nurse clini-cians, nurses, dentists, pharmacists, and clinic paraprofessionals; and health planning by public health nutritionists and public health administrators.

Means of nutritional assessment depend on goals, on availability of records, on time constraints, on the training of health personnel, on laboratory expertise, equipment, and facilities, and on the educational, psychiatric, and physical health status of the patients or clients.

Nutritional screening may include:

TABLE 16

**Common Diseases Which Change Food Energy or Nutrient Requirements**

| Disease | Change in nutrient requirement |
|---|---|
| Pneumonia | ↑ Food energy ↑ protein |
| Anemia | |
|    Iron deficiency | Iron ↑ Vitamin C ↑ |
|    Pernicious | Vitamin $B_{12}$↑ |
| Hypertension | ↓ Sodium |
| Congestive heart failure | ↓ Sodium ↑ Magnesium |
| Congestive heart failure | ↓ Sodium ↑ Magnesium |
| | ↑ Zinc |
| | ↑ Potassium |
| Coronary heart disease | ↓ Cholesterol |
| | ↓ Total fat |
| | ↓ Saturated |
| Alcoholic cirrhosis | B vitamins, Vitamin C, A, D ↑ |
| | Protein ↓ (if liver failure is present) |
| | Sodium ↓ |
| Chronic renal disease (not on dialysis) | Protein ↓ <br> Vitamin D ↑ |

1. Analysis of records
   a. Morbidity and mortality data (prevailing disease patterns)
   b. Incidence of anemia
   c. Incidence of obesity/underweight
   d. Incidence of low birth-weight infants (less than 2500 g)
   e. Incidence of dental caries
2. Nutritional history, including dietary questionnaire — abnormal findings suggesting a monotonous or bizarre diet would require further dietary assessment.
3. Medical history — information derived from the medical history may point to nutritional disease or to the risk of nutritional disease, e.g., alcoholism or any condition associated with loss of appetite, malabsorption, chronic infection, or cancer.
4. Drug history — information from the drug history will point to the risk of drug-induced nutritional deficiency or nutritional risks associated with drug addiction.
5. Pregnancy history — birth of one or more low birth-weight infants indicates high risk for pregnancy outcome in the future.
6. Physical examination — when signs suggesting nutritional disorder are present, it is necessary to carry out confirmatory tests including a biochemical assessment of nutritional status.
7. Anthropometry — nutritional assessment requires accurate measurement of height and weight or, in infants, length and weight, which can then be expressed as percent of standard weight/height/age. Additional measures include triceps skinfold thickness (mm), arm muscle circumference, and/or arm muscle mass, which are calculated from triceps skinfold thickness, and midupper arm circumference. Anthropometric measures are useful longitudinally to examine for changes in body weight or in muscle mass over time. Abnormal findings may require further workups including diagnostic studies.
8. Hematologic screening — includes determnation of hemoglobin and hematocrit. Abnormal values require a complete blood count as well as investigation of iron, folate, vitamin $B_{12}$ status, and, under specific conditions, vitamin $B_6$ status.

9. Dental screening — examination of the teeth should be carried out to define the number of decayed, missing, and filled teeth, as well as evidence of dental neglect. Presence of dental caries requires further nutritional evaluation and history-taking in regard to fluoride status and sugar in the diet.

## 1. Dietary Methodology and Its Utilization

Dietary assessment may be by (1) 24-hr recall of food intake, (2) food frequency, (3) diet diary, (4) diet history, (5) individual or household survey of purchased foods, and (6) actual food intake. The choice of method is dependent on whether the dietary assessment is of individuals or groups, and if of groups, on the characteristics and size of the sample. Quantitative dietary methods to determine nutrient intake include the 24-hr recall, the diet diary, the diet history, and observation of actual food consumed.[91]

Seven-day diet diaries are an excellent means of obtaining dietary information from cooperative patients who have been instructed in how to make these records. Less satisfactory are the 3- and 5-day diet diaries, but they may be used when there is reason to believe that the patient will lose interest in the record before the week is completed.

Determinations of actual food intake can be obtained by nurses. She/he monitors every item of food and drink consumed over a specified period, and then a replicate sample of the total food/beverage intake is prepared and subjected to nutrient analysis. An absolute requirement is that plate waste or tray waste is assessed and subtracted.

Diet histories are used by trained dietitians to obtain both quantitative and qualitative information. Nutrient intake is determined by analyzing diet records using food composition tables.[92-94]

Short dietary questionnaires are advocated for nutrition screening by nonnutritionist health professionals. They are intended to obtain information on dietary excesses as well as deficiencies. It is suggested that in nutrition screening, a food frequency record be obtained with quantitation of specific items and addition of information on eating patterns (Table 17). An alternate, short dietary questionnaire is shown in Table 18.

Household surveys may be used to determine the food and potential nutrient intake of families or in one-person households of individuals. Shopping lists can be used to assess the foods and hence the nutrients available to elderly housebound persons living alone.[96]

## 2. Clinical Methods

Signs of nutritional disease and particularly of nutritional deficiency may be elicited by physical examination. Composite signs of nutritional deficiency are summarized in Table 19. If a nutritional diagnosis is made from physical findings, then it must be supported by (1) biochemical assessment of nutritional status indicating one or more specific nutrient deficiencies and (2) the reversal or healing of the physical signs when therapeutic doses of the "missing" nutrients are administered by a route appropriate for optimal absorption.

## 3. Anthropometric Assessment
### a. Weight Standards

Scale measurements and mathematical expressions have been developed to define "underweight" and "overweight." These include:

1. Weight status in relationship to average weight for a given height, age, and sex
2. Ponderal index (Pi) which is defined as the height divided by the cube route of the weight

TABLE 17

**Food Frequency Questionnaire**

How many *times per week* do you consume

| | Circle correct number | More than 7 (specify) |
|---|---|---|

Chicken or turkey——————— 0  1  2  3  4  5  6  7  ————
Fish or tunafish——————— 0  1  2  3  4  5  6  7  ————
Hot dogs or cold cuts——————— 0  1  2  3  4  5  6  7  ————
Pork ——————— 0  1  2  3  4  5  6  7  ————
Liver——————— 0  1  2  3  4  5  6  7  ————
Other meats ——————— 0  1  2  3  4  5  6  7  ————
Cottage cheese——————— 0  1  2  3  4  5  6  7  ————
Cooked greens——————— 0  1  2  3  4  5  6  7  ————
Other cooked vegetables——————— 0  1  2  3  4  5  6  7  ————
Raw fruit ——————— 0  1  2  3  4  5  6  7  ————
Raw vegetables ——————— 0  1  2  3  4  5  6  7  ————
Dried cooked beans and peas——————— 0  1  2  3  4  5  6  7  ————
Instant breakfasts——————— 0  1  2  3  4  5  6  7  ————
Peanut butter——————— 0  1  2  3  4  5  6  7  ————
Potato chips or Fritos®——————— 0  1  2  3  4  5  6  7  ————
Crackers or pretzels——————— 0  1  2  3  4  5  6  7  ————
Macaroni, spaghetti, rice, noodles——— 0  1  2  3  4  5  6  7  ————
Mashed or baked potatoes——————— 0  1  2  3  4  5  6  7  ————
French fries——————— 0  1  2  3  4  5  6  7  ————
Candy bar ——————— 0  1  2  3  4  5  6  7  ————
Ice cream ——————— 0  1  2  3  4  5  6  7  ————
Cookies——————— 0  1  2  3  4  5  6  7  ————
Pie, cakes ——————— 0  1  2  3  4  5  6  7  ————
Doughnuts——————— 0  1  2  3  4  5  6  7  ————
Kool-Aid®——————— 0  1  2  3  4  5  6  7  ————
Soup——————— 0  1  2  3  4  5  6  7  ————
Yogurt ——————— 0  1  2  3  4  5  6  7  ————

How many servings per day do you eat or drink of the following:

Bread, toast, rolls, sweet rolls, muf-
fins (1 slice or 1 item is a serving)——— 0  1  2  3  4  5  6  7  ————
Milk (8 oz is a serving)——————— 0  1  2  3  4  5  6  7  ————
Butter or margarine (1 teaspoon is a
serving) ——————— 0  1  2  3  4  5  6  7  ————
Cheese (1 oz is a serving)——————— 0  1  2  3  4  5  6  7  ————
What kind do you usually
eat
Eggs (1 egg is a serving)——————— 0  1  2  3  4  5  6  7  ————
What form do you usually
eat
Coffee or tea (state) (1 cup is a serv-
ing)——————— 0  1  2  3  4  5  6  7  ————
What kind do you usually
drink
Soda (1 can is a serving)——————— 0  1  2  3  4  5  6  7  ————
What kind do you usually
drink
Cereal (3/4 cup to 1 cup is a serving)——— 0  1  2  3  4  5  6  7  ————
What brand do you usually
eat
Orange, grapefruit, pineapple, or to-
mato juice (4 oz is a serving)——— 0  1  2  3  4  5  6  7  ————

TABLE 17 (continued)

## Food Frequency Questionnaire

How many *times per week* do you consume

| | Circle correct number | More than 7 (specify) |
|---|---|---|

What kind do you usually drink

Beer (1 can is a serving)_____ 0  1  2  3  4  5  6  7 _____

What kind do you usually drink

Wine (1 glass is a serving)_____ 0  1  2  3  4  5  6  7 _____

What kind do you usually drink

Liquor (or mixed drink) (1 glass is a serving)_____ 0  1  2  3  4  5  6  7 _____

What kind do you usually drink

How many *times per week* do you eat the following meals

Breakfast _____ 0  1  2  3  4  5  6  7 _____
Lunch _____ 0  1  2  3  4  5  6  7 _____
Dinner_____ 0  1  2  3  4  5  6  7 _____

## TABLE 18

### Short Dietary Questionnaire

Do you drink milk?                    Yes ☐        No ☐

If yes, whole milk ☐
2% Milk ☐
Skim milk ☐
Other ☐
Specify_____

Please indicate which of the following foods you eat and how often.

| | Never or hardly ever (less than once a week) | Sometimes (not daily but at least once a week) | Every day or nearly every day |
|---|---|---|---|
| Cheese, yogurt, ice cream | ☐ | ☐ | ☐ |
| Eggs | ☐ | ☐ | ☐ |
| Dried beans, peas, peanut butter | ☐ | ☐ | ☐ |
| Meat, fish, poultry | ☐ | ☐ | ☐ |
| Bread, rice, pasta, grits, cereal, tortillas, potatoes | ☐ | ☐ | ☐ |
| Fruits or fruit juices | ☐ | ☐ | ☐ |
| Vegetables | ☐ | ☐ | ☐ |

If you eat fruits or drink fruit juices every day or nearly every day, which ones do you eat or drink most often? (Not more than three.)

_____

_____

TABLE 18 (continued)

**Short Dietary Questionnaire**

If you eat vegetables every day or nearly every day, which ones do you eat most often? (Not more than three.)

_____

_____

|  | Yes | No |
|---|---|---|
| Do you usually eat anything between meals? | ☐ | ☐ |

    If yes, name the two or three snacks
     (including bedtime snacks) that
     you have most often.

_____

_____

Do you or the person who prepares your meals have use of a

|  | Yes | No |
|---|---|---|
|   Working stove | ☐ | ☐ |
|   Refrigerator | ☐ | ☐ |
|   Piped water | ☐ | ☐ |
| Do you take vitamins or iron? | ☐ | ☐ |

  If yes, how often? _____
  What kind? _____

| Are you on a special diet? | ☐ | ☐ |
|---|---|---|

  If yes, what is the reason?
    Allergy; specify type of diet ☐

_____

    Weight reduction; specify type of
    diet
    Other; specify reason for diet and
     type of diet     ☐

_____

Who recommended the diet? _____

Do you eat clay, paint chips, or anything else not usually considered food?
  If yes, what? _____
  How often? _____

From Fomon, S. J., *Nutritional Disorders of Children* — Screening, Followup, Prevention, Bureau of Community Health, Department of Health, Education and Welfare, Publ. No. (HSA) 76-5612, 1976, 10, and from Preliminary Guide for Developing Nutrition Services in Health Care Programs, U.S. Department of Health, Education and Welfare, Public Health Services, Health Services Administration, Bureau of Community Health Services, Rockville, Md., 1976.

## TABLE 19

### Clinical Signs of Nutritional Disease

| Deficiency | Skin | Mucosa | Bones/ joints | Blood | Brain | Peripheral nerves | Heart | Muscle/ kidney/ liver | Eyes | ECF* |
|---|---|---|---|---|---|---|---|---|---|---|
| Vitamins | | | | | | | | | | |
| A | Phrynoderma | Infections | — | — | — | — | — | — | Xerophthalmia, keratomalacia | — |
| B$_{12}$ | — | Glossitis | — | Megaloblastic anemia | Organic brain syndrome | Neuritis | — | — | — | — |
| D | — | — | Rickets, osteomalacia | — | — | — | — | Weakness | — | — |
| K | Purpura | — | — | Hemorrhage | — | — | — | — | — | — |
| E | — | — | — | Hemolytic anemia | — | — | — | — | — | Edema |
| C (scurvy) | Purpura | Bleeding gums | Joint hemorrhage | Anemia | — | — | — | — | — | — |
| B$_1$ | — | — | — | — | Psychosis (Wernicke) | Neuritis | Beri-beri heart | — | — | Edema |
| Riboflavin (B$_2$) | Dermatitis | Glossitis | — | — | — | — | — | — | Photophobia, conjunctivitis | — |
| Folic acid | — | Glossitis | — | Megaloblastic anemia | Organic brain syndrome | — | — | — | — | — |
| B$_6$ | Dermatitis | Glossitis | — | — | Organic brain syndrome | Neuritis | — | — | — | — |

TABLE 19 (continued)

**Clinical Signs of Nutritional Disease**

| Deficiency | Skin | Mucosa | Bones/joints | Blood | Brain | Peripheral nerves | Heart | Muscle/kidney/liver | Eyes | ECF[a] |
|---|---|---|---|---|---|---|---|---|---|---|
| Vitamins | | | | | | | | | | |
| B$_{12}$ | — | Glossitis | — | Megaloblastic anemia | Organic brain syndrome | Neuritis | — | — | — | — |
| Niacin (Pellagra) | Dermatitis | Glossitis, Enteritis | — | — | Organic brain syndrome | — | — | — | — | — |
| EFA[b] | Dermatitis | — | — | — | — | — | — | — | — | — |
| Protein | Flaky paint[c] dermatitis | — | Growth[c] failure | — | — | — | — | Fatty liver | — | Edema |
| Zinc | Impaired wound healing; exfoliative dermatitis | Loss of taste | Growth failure | — | — | — | — | — | — | — |
| Iron | — | Precancerous changes | — | Microcytic anemia | — | — | Cardiac failure | — | — | — |

[a]   ECF = extracellular fluid.
[b]   EFA = essential fatty acid.
[c]   Infants and children only.

# TABLE 20

## Ideal Weight for Height Charts[a]

| Height (cm) | Weight (kg) | Height (cm) | Weight (kg) | Height (cm) | Weight (kg) |
|---|---|---|---|---|---|
| | | **Ideal Weight for Males** | | | |
| 157 | 58.6 | 167 | 64.6 | 177 | 71.6 |
| 158 | 59.3 | 168 | 65.2 | 178 | 72.4 |
| 159 | 59.9 | 168 | 65.9 | 179 | 73.3 |
| 160 | 60.5 | 170 | 66.6 | 180 | 74.2 |
| 161 | 61.1 | 171 | 67.3 | 181 | 75.0 |
| 162 | 61.7 | 172 | 68.0 | 182 | 75.8 |
| 163 | 62.3 | 173 | 68.7 | 183 | 76.5 |
| 164 | 62.9 | 174 | 69.4 | 184 | 77.3 |
| 165 | 63.5 | 175 | 70.1 | 185 | 78.1 |
| 166 | 64.0 | 176 | 70.8 | 186 | 78.9 |
| | | **Ideal Weight for Females** | | | |
| 140 | 44.9 | 150 | 50.4 | 160 | 56.2 |
| 141 | 45.4 | 151 | 51.0 | 161 | 56.9 |
| 142 | 45.9 | 152 | 51.5 | 162 | 57.6 |
| 143 | 46.4 | 153 | 52.0 | 163 | 58.3 |
| 144 | 47.0 | 154 | 52.5 | 164 | 58.9 |
| 145 | 47.5 | 155 | 53.1 | 165 | 59.5 |
| 146 | 48.0 | 156 | 53.7 | 166 | 60.1 |
| 147 | 48.6 | 157 | 54.3 | 167 | 60.7 |
| 148 | 49.2 | 158 | 54.9 | 168 | 61.4 |
| 149 | 49.8 | 159 | 55.5 | 169 | 62.1 |

[a] These tables correct the 1959 Metropolitan Standards to nude weight.

References used were Jelliffe, D. B., The Assessment of the Nutritional Status of the Community, WHO Monograph No. 53, World Health Organization, Geneva, 1966, and Society of Actuaries, Build and Blood Pressure Study, Chicago, 1959.

3. Weight divided by height (W/H)
4. Weight divided by height squared (W/H$^2$)

Weight standards for given heights are derived from National Health Survey tables[97] and the Build and Blood Pressure Study.[98] Metropolitan Life Insurance height/weight tables based on the Build and Blood Pressure Study are commonly used, defining desirable weight, underweight, and overweight (Table 20).[99]

### b. Triceps Skinfold Thickness

Fatfold measurements are easily made using Lange calibrated skinfold calipers. Triceps skinfold thickness is an excellent measure of fatness in both sexes. Standards for triceps skinfold thickness (mm) for both sexes and different ages are those of Mayer and Seltzer (Table 21).[100]

### c. Arm Circumference and Arm-Muscle Circumference

Measurement of midupper arm circumference can be used to roughly determine thin-

TABLE 21

| Age (years) | Skinfold measurements | |
| --- | --- | --- |
| | Males | Females |
| 12 | 18 | 22 |
| 13 | 18 | 23 |
| 14 | 17 | 23 |
| 15 | 16 | 24 |
| 16 | 15 | 25 |
| 17 | 14 | 26 |
| 18 | 15 | 27 |
| 19 | 15 | 27 |
| 20 | 16 | 28 |
| 21 | 17 | 28 |
| 22 | 18 | 28 |
| 23 | 18 | 28 |
| 24 | 19 | 28 |
| 25 | 20 | 29 |
| 26 | 20 | 29 |
| 27 | 21 | 29 |
| 28 | 22 | 29 |
| 29 | 23 | 29 |
| 30—50 | 23 | 30 |

Adapted from Seltzer, C. C. and Mayer, J., *Postgrad. Med.*, 38, 101, 1965. With permission.

ness or fatness. Used in conjunction with triceps skinfold thickness, it can be used to determine arm muscle circumference, which is a measure of lean body mass. Arm muscle circumference (cm) = arm circumference (cm) = [π × triceps skinfold (mm)]. Standards for arm muscle circumference have been developed by Frisancho (Table 22).[101]

Serial measurements of triceps skinfold thickness and arm muscle circumference provide an acceptable profile of change in body fatness and muscle mass.

### 4. Hematological Assessment

Measurements of hemoglobin and hematocrit have routinely been used in screening for anemia. It has recently been shown by Garn et al. that ethnic differences exist in hemoglobin values, such that blacks have been found to have normal hemoglobin levels lower than whites.[102] From the nutritional standpoint, broad definition of anemias into microcytic (iron deficiency) and macrocytic (folacin or vitamin $B_{12}$ deficiency) can be obtained from the complete blood count. Normal blood counts and values indicative of nutritional anemias are shown in Table 23. Nutritional diagnosis of anemia also requires laboratory determinaion of iron, vitamin $B_{12}$, and folacin status.

High hemoglobin values as well as high red blood cell counts are associated with gross obesity as well as with chronic bronchitis or other lung diseases.

### 5. Biochemical Assessment

Biochemical measurements used to determine protein status include serum albumin and/or serum transferrin.[103] Diagnosis of iron deficiency is best made on the basis of

43

serum iron, free erythrocyte protoporphyrin, and serum ferritin levels. Elevation of free erythrocyte protoporphyrin occurs in iron deficiency but is also found in lead toxicity. In the assessment of vitamin nutriture, biochemical tests are of the utmost importance. Levels of water-soluble vitamins in the plasma reflect recent dietary intakes. Red blood cell levels of vitamins such as folacin are more valid measures of vitamin status.

Enzyme tests of vitamin status have been developed as indicators of vitamin function, including the erythrocyte transketolase test for thiamin status and the erythrocyte glutathione reductase assay for riboflavin status. Acceptable, low, and deficient laboratory values for key nutrients are shown in Table 24.

## 6. Immunological Assessment

Cellular immune function is impaired in adults as well as children with protein or protein-energy malnutrition. Immunological screening is by total lymphocyte count and by skin tests for delayed hypersensitivity reactions.[104]

TABLE 22

Percentiles for Upper Arm Diameter and Upper Arm Circumference for Whites of the Ten-State Nutrition Survey of 1968 to 1970

| Age midpoint, years | Arm muscle | | | | | |
|---|---|---|---|---|---|---|
| | Diameter percentiles (mm) | | | Circumference percentiles (mm) | | |
| | 5th | 50th | 85th | 5th | 50th | 95th |
| **Males** | | | | | | |
| 0.3 | 26 | 34 | 40 | 81 | 106 | 133 |
| 1 | 32 | 39 | 44 | 100 | 123 | 146 |
| 2 | 35 | 40 | 44 | 111 | 127 | 146 |
| 3 | 36 | 42 | 46 | 114 | 132 | 152 |
| 4 | 38 | 43 | 48 | 118 | 135 | 157 |
| 5 | 39 | 45 | 50 | 121 | 141 | 166 |
| 6 | 40 | 47 | 51 | 127 | 146 | 167 |
| 7 | 41 | 48 | 52 | 130 | 151 | 173 |
| 8 | 44 | 50 | 55 | 138 | 158 | 185 |
| 9 | 44 | 51 | 58 | 138 | 161 | 200 |
| 10 | 45 | 53 | 59 | 142 | 168 | 202 |
| 11 | 48 | 55 | 62 | 150 | 174 | 211 |
| 12 | 49 | 58 | 66 | 153 | 181 | 221 |
| 13 | 51 | 62 | 71 | 159 | 195 | 242 |
| 14 | 53 | 67 | 74 | 167 | 211 | 265 |
| 15 | 55 | 70 | 80 | 173 | 220 | 271 |
| 16 | 59 | 73 | 83 | 186 | 229 | 281 |
| 17 | 66 | 78 | 86 | 206 | 245 | 290 |
| 21 | 69 | 82 | 91 | 217 | 258 | 305 |
| 30 | 70 | 86 | 94 | 220 | 270 | 315 |
| 40 | 71 | 86 | 96 | 222 | 270 | 318 |
| **Females** | | | | | | |
| 0.3 | 27 | 33 | 37 | 86 | 104 | 126 |
| 1 | 31 | 37 | 41 | 97 | 117 | 135 |
| 2 | 34 | 40 | 44 | 105 | 125 | 146 |

TABLE 22 (continued)

**Percentiles for Upper Arm Diameter and Upper Arm Circumference for Whites of the Ten-State Nutrition Survey of 1968 to 1970**

Arm muscle

| Age midpoint, years | Diameter percentiles (mm) | | | Circumference percentiles (mm) | | |
|---|---|---|---|---|---|---|
| | 5th | 50th | 85th | 5th | 50th | 95th |
| | | | Females | | | |
| 3 | 34 | 41 | 44 | 108 | 128 | 143 |
| 4 | 36 | 42 | 46 | 114 | 132 | 152 |
| 5 | 38 | 44 | 48 | 119 | 138 | 160 |
| 6 | 38 | 45 | 49 | 121 | 140 | 165 |
| 7 | 39 | 47 | 52 | 123 | 146 | 175 |
| 8 | 41 | 48 | 53 | 129 | 151 | 186 |
| 9 | 43 | 50 | 56 | 136 | 157 | 193 |
| 10 | 44 | 52 | 58 | 139 | 163 | 196 |
| 11 | 44 | 55 | 62 | 140 | 171 | 209 |
| 12 | 48 | 57 | 64 | 150 | 179 | 212 |
| 13 | 49 | 59 | 66 | 155 | 185 | 225 |
| 14 | 53 | 61 | 70 | 166 | 193 | 234 |
| 15 | 52 | 62 | 70 | 163 | 195 | 232 |
| 16 | 54 | 64 | 72 | 171 | 200 | 260 |
| 17 | 54 | 62 | 71 | 171 | 196 | 241 |
| 21 | 54 | 65 | 73 | 170 | 205 | 253 |
| 30 | 56 | 68 | 78 | 177 | 213 | 272 |
| 40 | 57 | 69 | 80 | 180 | 216 | 279 |

From Frisancho, A. R., Triceps skin fold and upper arm muscle size norms for assessment of nutritional status, *Am. J. Clin. Nutr.*, 27, 1052, 1974. With permission.

TABLE 23

**Guidelines Used for the Interpretation of Blood Data***

Determination

| Age (years) | Sex | Hemoglobin (g/100 ml) | | | Hematocrit (%) | | |
|---|---|---|---|---|---|---|---|
| | | Deficient | Low | Acceptable | Deficient | Low | Acceptable |
| < 2 | M-F | < 9.0 | 9.0—9.9 | ≥ 10.0 | < 28 | 28—30 | ≥ 31 |
| 2—5 | M-F | < 10.0 | 10.0—10.9 | ≥ 11.0 | < 30 | 30—33 | ≥ 34 |
| 6—12 | M-F | < 10.0 | 10.0—11.4 | ≥ 11.5 | < 30 | 30—35 | ≥ 36 |
| 13—16 | M | < 12.0 | 12.0—12.9 | ≥ 13.0 | < 37 | 37—39 | ≥ 40 |
| | F | < 10.0 | 10.0—11.4 | ≥ 11.5 | < 31 | 31—35 | ≥ 36 |
| > 16 | M | < 12.0 | 12.0—13.9 | ≥ 14.0 | < 37 | 37—43 | ≥ 44 |
| | F | < 10.0 | 10.0—11.9 | ≥ 12.0 | < 31 | 31—37 | ≥ 38 |
| Pregnant 2nd trimester | | < 9.5 | 9.5—10.9 | ≥ 11.0 | < 30 | 30—34 | ≥ 35 |
| 3rd trimester | | < 9.0 | 9.0—10.4 | ≥ 10.5 | < 30 | 30—32 | ≥ 33 |

M = males    F = females

* Adapted from the reports of O'Neal et al. and of the Ten-State Nutrition Survey.

TABLE 24

Interpretation of Laboratory Tests for the Evaluation of Nutritional Status (Adults)

| Test | Deficient (high risk) | Low (medium risk) | Acceptable (low risk) | Nutrient evaluation |
|---|---|---|---|---|
| Serum iron $\mu$g/100 m$l$ | < 30 | 30—59.9 | $\geqslant$ 60 | Iron |
| Serum albumin g/100 m$l$ | < 2.8 | 2.8—3.4 | $\geqslant$ 3.5 | Protein |
| Plasma carotene $\mu$g/100 m$l$ | < 20 | 20—39 | $\geqslant$ 40 | GI absorption |
| Plasma retinol $\mu$g/100 m$l$ | < 10 | 10—19 | $\geqslant$ 20 | Vitamin A |
| Serum folate ng/m$l$ | < 3 | 3.0—5.9 | > 6.0 | Folate |
| Erythrocyte folate ng/m$l$ | < 140 | 140—159 | > 160 | Folate |
| Serum vitamin $B_{12}$ pg/m$l$ | < 150 | 150—200 | > 200 | Vitamin $B_{12}$ |
| Serum ascorbic acid mg/ 100 m$l$ | < 0.20 | 0.20—0.29 | $\geqslant$ 0.30 | Vitamin C |
| Erythrocyte transketolase % (thiamin pyrophosphate stimulation) | > 20 | 16—20 | 0—15 | Thiamin |
| Erythrocyte glutathione reductase-activity coefficients | > 1.40 | 1.20— 1.40 | < 1.20 | Riboflavin |
| Erythrocyte transaminase indices | | | | |
| EGPT | > 1.25 | — | $\leqslant$ 1.25 | Vitamin $B_6$ |
| EGOT | > 2.0 | — | $\leqslant$ 2.0 | Vitamin $B_6$ |
| Tryptophan load test increase in excretion xanthurenic acid mg/day | > 50 | 25—50 | < 25 | Vitamin $B_6$ |
| $N^1$-methyl nicotinamide excretion mg/g creatinine | < 0.5 | 0.5—1.59 | 1.6—4.29 | Niacin |
| Serum CA × P product mg% | < 30 | — | > 30 | Vitamin D |
| Alkaline phosphatase King-Armstrong units/100 m$l$ | >20 (Adult) > 30 (Child) | — | 8—14 | Vitamin D |

From Roe, D. A., *Drug-induced Nutritional Deficiencies*, AVI Press, Westport, Conn., 1976. With permission.

# SECTION III
# ABUSE OF FOOD SUBSTANCE

## A. Obesity: Causes, Treatment, and Prevention

### 1. Causes

The balance between food intake and energy expenditure determines nutrient storage as fat. The fat of the body is stored in the fat layer under the skin as well as around and between deeper organs. It is stored in fat cells or adipocytes. In obesity there is an increase in the number of fat cells and the size of fat cells or, more commonly, in both the number and size of these cells. The number of fat cells increases from the neonatal period up to the time of puberty. In early infancy, overfeeding causes an excessive increase in the number of fat cells, and this may also occur during childhood. Once these fat cells have developed, it appears that they usually persist. There is evidence that an excessive number of fat cells predisposes children to develop obesity either in later childhood or in adolescence, and that this obesity will persist in adult life. While many writers and investigators have come to the conclusion that prevention of the development of an excessive number of fat cells can be brought about by avoidance of overfeeding in early life, the generalization that fat babies become fat adults has been disputed. A more generally accepted conclusion is that fat people who have been obese in childhood or adolescence are more difficult to treat than those who have a later onset of obesity.[105]

Garn et al. have reported that obesity is associated with ethnic and social factors.[106] Children of obese parents tend to be fatter than children of lean parents. In the U.S. and in other industrialized countries, obesity is commoner in less educated women and is most prevalent in socially isolated women.[107]

While there are rare hereditary obesities, evidence points to an association between familial obesity and familial excessive intake of food. Indeed, large intake of food as a family custom may be equated with positive health and may be culturally acceptable within groups of people who are less socially mobile. Overeating with secondary obesity may also be related to anxiety, lack of self-esteem, or boredom. Fat people tend to be sedentary, avoiding exercise, partly perhaps because they feel themselves ill-adapted to sustained activity and partly because it is their innate characteristic to avoid vigorous movement.

Body fat is derived from dietary sources including dietary fats and carbohydrates. The fat that is deposited in the subcutaneous areas of the body acts not only as insulation against cold, but also as an energy store.

Obesity is rarely associated with endocrine abnormalities. The public has been misled in the idea that an "underactive" thyroid or pituitary gland is the usual cause of obesity.

### a. Complications and Consequences

Obesity is associated with the development of maturity-onset diabetes[108] and hypertension.[109] In very fat people, dermatitis with secondary yeast infection is common between fat folds. Obesity is also a risk factor in the development of heart disease.[110] While obesity does not cause back problems, arthritis, varicose veins, or abdominal hernias, disability associated with these conditions and more severe symptomatology is associated with the presence of obesity. Symptoms of gall bladder disease arise more frequently in fat people, probably because of high intakes of fatty foods, though the causal relationship has not been fully established. Very fat people are at greater risk in surgical procedures, and may develop postoperative respiratory complications because of decreased vital capacity.[111,112]

In job situations, fat people tend to be less agile, to be intolerant of heat, and to be less efficient in occupations requiring sustained physical activity.[113]

Unemployment is associated with obesity, particularly in women. However, evidence points to a dual association in that whereas employers may be less inclined to hire fat people, lack of occupation may lead to a situation of boredom, and hence overeating. The association between obesity and lack of a job is also conditioned by the fact that complications of obesity may reduce employability.[114]

## 2. Treatment

Treatment of obesity should be preceded by comprehensive medical screening to establish whether or not there are any physical or mental contraindications to prescription of a weight-reducing regime. This screening process should include the following components:

### a. Medical History

The presence or recent history of any of the following diseases should be recorded:

1.  Diseases requiring special diets: diabetes mellitus, peptic ulcer, gall bladder disease, gout
2.  Diseases that preclude or limit physical activity: rheumatic heart disease, coronary artery disease (heart attacks), arthritis, anemia, chronic neurological diseases (e.g., multiple sclerosis)
3.  Diseases that will mitigate against success of therapy: alcoholism, drug abuse, schizophrenia, depression (including hospitalization for psychiatric illness), and mental retardation

### b. Physical Examination

Examination should be for evidence of hypertension (check blood pressure on at least three occasions, using large sphygmomanometer cuff), edema, dermatitis, arthritis, varicose veins, deformities, and lack of mobility.

### c. Ancillary Tests

Tests should include a blood count to check for anemia, an EKG examination for cardiac arythmias, urinalysis for glycosuria, and fasting biochemical profile to assess blood sugar, as well as liver and kidney function. It is also advisable to perform thyroid function tests ($T_3$ and $T_4$) though test values may be outside normal limits if the patient has been receiving thyroid medication.

The components of acceptable therapy for obesity are (1) diet, (2) exercise, and (3) behavior modification. We advocate a combination of these modalities for most patients. The patient should always assist in the initial decision-making process as to the amount of weight loss that is desired. If a nutritionist is not available, it is the responsibility of the nurse or physician to explain how long it is likely to take to achieve the given weight loss, given adherence to the prescribed diet, with or without addition of an exercise plan.

Prediction of weight loss requires first that the food-energy intake needed to maintain body weight is ascertained. The patient should be asked to eat as usual and record all food that is consumed. It is important to explain to the patient that during this period, there should be no food restriction, and that the amount of physical activity should be held constant. Food intake should be recorded for 7 days. From this record, food-energy intake per day can be calculated by taking energy intake for 1 week and dividing by seven (Table 25). It is assumed that during the period of weight reduction,

TABLE 25

Calculation of Food Energy Intake. Nutritive Value of Common Foods[a]

| | Amount | Cal | Protein (g) | Fat (g) | Carbohydrate (g) | Calcium (mg) | Phosphorus (mg) | Iron (mg) | Sodium (mg) | Potassium (mg) | Vitamin A (IU) | Thiamin (mg) | Riboflavin (mg) | Niacin (mg) | Ascorbic acid (mg) | USDA Handbook No. 456/ Ref. no. |
|---|---|---|---|---|---|---|---|---|---|---|---|---|---|---|---|---|
| **Milk** | | | | | | | | | | | | | | | | |
| Buttermilk | 1C | 88 | 8.8 | 0.2 | 12.5 | 296 | 233 | 0.1 | 319 | 343 | 10 | 0.10 | 0.44 | 0.2 | 2 | 509b |
| Milk, skim | 1C | 88 | 8.8 | 0.2 | 12.5 | 296 | 233 | 0.1 | 127 | 355 | 10 | 0.09 | 0.44 | 0.2 | 2 | 1322b |
| Milk, whole | 1C | 159 | 8.5 | 8.5 | 12.0 | 288 | 227 | 0.1 | 122 | 351 | 350 | 0.07 | 0.41 | 0.2 | 2 | 1320b |
| **Vegetables** | | | | | | | | | | | | | | | | |
| Asparagus[b] | 4 | | | | | | | | | | | | | | | 47b |
| Green beans, frozen, french style | ½C | 14 | 1.2 | 0.1 | 3.1 | 19 | 25 | 0.5 | 1 | 99 | 443 | 0.07 | 0.08 | 0.5 | 10 | 194c |
| Beans, lima (frozen) | ½C | 84 | 5.1 | 0.1 | 16.2 | 17 | 77 | 1.5 | 86 | 362 | 195 | 0.06 | 0.03 | 0.9 | 15 | 173c |
| Beets | ½C | 27 | 1.0 | 0.1 | 6.1 | 12 | 20 | 0.5 | 37 | 177 | 15 | 0.03 | 0.04 | 0.3 | 5 | 385b |
| Beet greens,[b] spinach, collards (cooked) | ½C | 22 | 2.5 | 0.4 | 3.5 | 111 | 34 | 1.4 | 33 | 260 | 6133 | 0.07 | 0.14 | 0.6 | 36 | 393a, 2170a, 807a |
| Brussel sprouts,[b] frozen | ½C | | | | | | | | | | | | | | | 492c |
| Cabbage, raw | 1C | 21 | 1.7 | 0.2 | 4.2 | 25 | 34 | 0.5 | 12 | 189 | 200 | 0.06 | 0.06 | 0.4 | 47 | 512c |
| Cauliflower | ½C | | | | | | | | | | | | | | | 831a |
| Corn on cob, 5" (boiled) | 1 | 70 | 2.5 | 0.8 | 16.2 | 2 | 69 | 0.5 | —[c] | 151 | 310 | 0.09 | 0.08 | 1.1 | 7 | 846a |
| Peas, green, frozen | ½C | 55 | 4.1 | 0.3 | 9.5 | 15 | 69 | 1.5 | 92 | 108 | 480 | 0.22 | 0.07 | 1.4 | 11 | 1530c |
| Potato, french fried (2" to 3½" diameter) | 10 | 137 | 2.2 | 6.6 | 18.0 | 8 | 56 | 0.7 | 3 | 427 | — | 0.07 | 0.04 | 1.6 | 11 | 1789b |
| Potato, sweet, mashed | ¼C | 73 | 1.1 | 0.3 | 16.8 | 21 | 30 | 0.5 | 7 | 155 | 5038 | 0.06 | 0.04 | 0.4 | 11 | 2250c |
| Potato, white, baked (2" diameter) | ½C | 59 | 1.7 | 0.1 | 13.3 | 6 | 41 | 0.5 | 3 | 316 | — | 0.07 | 0.03 | 1.2 | 13 | 1787d |

TABLE 25 (continued)

Calculation of Food Energy Intake. Nutritive Value of Common Foods[*]

| | Amount | Cal | Pro-tein (g) | Fat (g) | Carbo-hydrate (g) | Cal-cium (mg) | Phos-phorus (mg) | Iron (mg) | Sodium (mg) | Potas-sium (mg) | Vitamin A (IU) | Thiamin (mg) | Ribo-flavin (mg) | Niacin (mg) | Ascor-bic acid (mg) | USDA Hand-book No. 456/ Ref. no. |
|---|---|---|---|---|---|---|---|---|---|---|---|---|---|---|---|---|
| **Vegetables (continued)** | | | | | | | | | | | | | | | | |
| Pumpkin, carrots, winter squash | ½C | 37 | 1.1 | 0.3 | 89 | 27 | 32 | 0.5 | 10 | 261 | 6757 | 0.04 | 0.08 | 0.6 | 7 | 1832d, 620a, 2201a |
| Rutabagas, mashed | ½C | 42 | 1.1 | 0.1 | 9.9 | 71 | 37 | 0.4 | 5 | 201 | 660 | 0.07 | 0.07 | 1.0 | 31 | 1920b |
| Summer squash | ½C | | | | | | | | | | | | | | | 2192a |
| Celery, stalks | 3 | | | | | | | | | | | | | | | 637c |
| Cucumbers, slices | 16 | 9 | 0.7 | 0.1 | 2.1 | 19 | 17 | 0.6 | 18 | 133 | 290 | 0.03 | 0.04 | 0.3 | 6 | 942d |
| Lettuce, shredded | 1C | | | | | | | | | | | | | | | 1256c |
| Tomatoes, canned and juice | ½C | 34 | 1.2 | 0.2 | 5.2 | 8 | 23 | 0.9 | 200 | 269 | 1028 | 0.06 | 0.04 | 0.9 | 20 | 2284d, 2288d |
| Tomatoes, fresh (2 2/5" diameter) | 1 | 20 | 1.0 | 0.2 | 4.3 | 12 | 25 | 0.5 | 3 | 222 | 820 | 0.05 | 0.04 | 0.6 | 21 | 2282c |
| Turnips, mashed | ½C | 26 | 0.9 | 0.3 | 5.7 | 41 | 28 | 0.5 | 39 | 216 | — | 0.05 | 0.06 | 0.4 | 26 | 2353b |
| **Fruits** | | | | | | | | | | | | | | | | |
| Apple (2½" diameter) | 1 | 51 | 0.2 | 0.6 | 15.3 | 7 | 11 | 0.3 | 1 | 116 | 100 | 0.03 | 0.02 | 0.1 | 4 | 13d |
| Applesauce, unsweetened | ½C | 50 | 0.3 | 0.3 | 13.2 | 5 | 6 | 0.6 | 3 | 95 | 50 | 0.03 | 0.01 | 0.1 | 1 | 28e |
| Banana, medium | ½ | 51 | 0.7 | 0.1 | 13.2 | 5 | 16 | 0.4 | 1 | 220 | 115 | 0.03 | 0.04 | 0.4 | 6 | 141b |
| Berries (blue, black, raspberries) | ½C | 46 | 0.8 | 0.4 | 10.9 | 19 | 15 | 0.8 | 1 | 113 | 118 | 0.02 | 0.04 | 0.4 | 10 | 424b, 418a, 1851a |
| Cantaloupe (¼ of 6" diameter) | 1C | 48 | 1.1 | 0.2 | 12.0 | 22 | 26 | 0.6 | 19 | 402 | 5440 | 0.06 | 0.05 | 1.0 | 53 | 1358c |

## Fruits (continued)

| Food | Amount | | | | | | | | | | | | | | | Code |
|---|---|---|---|---|---|---|---|---|---|---|---|---|---|---|---|---|
| Citrus fruit: | | | | | | | | | | | | | | | | |
| Orange*/grapefruit juice | ½C | | | | | | | | | | | | | | | 1421c, 1053b |
| Orange, small | 1 | 49 | 0.8 | 0.1 | 11.8 | 19 | 19 | 0.3 | 1 | 189 | 158 | 0.08 | 0.3 | 0.3 | 49 | 1437b, 1061a |
| Grapefruit | ½ | | | | | | | | | | | | | | | |
| Cherries, large | 10 | 47 | 0.9 | 0.2 | 11.7 | 15 | 13 | 0.3 | 1 | 129 | 70 | 0.03 | 0.04 | 0.3 | 7 | 663b |
| Grapes | 12 | 41 | 0.4 | 0.2 | 10.4 | 7 | 12 | 0.2 | 2 | 104 | 60 | 0.04 | 0.02 | 0.2 | 2 | 1085a |
| Pears, water pack | 1 | 50 | 0.4 | 0.4 | 1.8 | 8 | 10 | 0.4 | 2 | 136 | — | 0.02 | 0.04 | 0.2 | 2 | 1504e |
| Raisins (1½ T) | ½ oz. | 40 | 0.4 | — | 10.8 | 9 | 14 | 0.5 | 4 | 107 | — | 0.02 | 0.01 | 0.1 | — | 1846c |
| Strawberries | 1C | 55 | 1.0 | 0.7 | 12.5 | 31 | 31 | 1.5 | 1 | 244 | 90 | 0.04 | 0.10 | 0.9 | 88 | 2217e |
| Watermelon, diced | 1C | 42 | 0.8 | 0.3 | 10.2 | 11 | 16 | 0.8 | 2 | 160 | 940 | 0.05 | 0.05 | 0.3 | 11 | 2424c |

## Bread and Crackers, Enriched

| Food | Amount | | | | | | | | | | | | | | | Code |
|---|---|---|---|---|---|---|---|---|---|---|---|---|---|---|---|---|
| Biscuit (2" diameter, 1¼" high) | 1 | 103 | 2.1 | 4.8 | 12.8 | 34 | 49 | 0.4 | 175 | 33 | — | 0.06 | 0.06 | 0.5 | — | 410a |
| Cornbread, piece (2½"x2½"x1½") | 1 | 178 | 3.8 | 5.8 | 27.5 | 133 | 209 | 0.8 | 263 | 61 | 130 | 0.10 | 0.10 | 0.8 | — | 1350d |
| Frankfurter roll (6") | 1 | 119 | 3.3 | 2.2 | 21.2 | 30 | 34 | 0.8 | 202 | 38 | — | 0.11 | 0.07 | 0.9 | — | 1902c |
| Graham crackers | 2 | 55 | 1.1 | 1.3 | 10.4 | 6 | 21 | 0.2 | 95 | 55 | — | 0.01 | 0.03 | 0.2 | — | 914b |
| Hamburger bun (3½") | 1 | 119 | 3.3 | 2.2 | 21.2 | 30 | 34 | 0.8 | 202 | 38 | — | 0.11 | 0.07 | 0.9 | — | 1902c |
| Muffin, plain (2"x1½") | 1 | 118 | 3.1 | 4.0 | 16.9 | 42 | 60 | 0.6 | 176 | 50 | 40 | 0.07 | 0.09 | 0.6 | — | 1343b |
| Saltines (2½" square) | 4 | 48 | 1.0 | 1.3 | 8.0 | 2 | 10 | 0.1 | 123 | 13 | — | — | — | 0.1 | — | 916d |
| Soda crackers (2½" square) | 5 | 65 | 1.3 | 1.8 | 10.0 | 3 | 13 | 0.2 | 156 | 16 | — | — | — | 0.1 | — | 918d |
| White or whole wheat bread | 1 | 76 | 2.4 | 0.9 | 14.1 | 24 | 27 | 0.7 | 142 | 29 | — | 0.07 | 0.06 | 0.7 | — | 461b |

## Cereals, Enriched

| Food | Amount | | | | | | | | | | | | | | | Code |
|---|---|---|---|---|---|---|---|---|---|---|---|---|---|---|---|---|
| Bran flakes, 40% | ½C | 53 | 1.8 | 0.3 | 14.1 | 10 | 63 | 6.2 | 104 | 68 | 825 | 0.20 | 0.25 | 2.1 | 6 | 441 |
| Corn flakes | ¾C | 73 | 1.5 | 0.1 | 15.9 | — | 7 | 0.4 | 188 | 22 | 885 | 0.22 | 0.26 | 2.2 | 7 | 866a |
| Farina, cooked | ½C | 51 | 1.6 | 0.1 | 10.6 | 5 | 15 | — | 176 | 11 | — | 0.05 | 0.03 | 0.5 | — | 992 |
| Grits, corn | ¾C | 62 | 1.4 | 0.1 | 13.5 | 1 | 13 | 0.4 | 251 | 14 | 75 | 0.05 | 0.04 | 0.5 | — | 863a |
| Oatmeal, cooked | ½C | 66 | 2.4 | 1.2 | 11.6 | 11 | 68 | 0.7 | 262 | 73 | — | 0.09 | 0.02 | 0.1 | — | 1391 |
| Rice, cooked | ½C | 74 | 1.4 | 0.1 | 16.4 | 7 | 19 | 0.6 | 256 | 19 | — | 0.08 | 0.01 | 0.7 | — | 1872a |
| Wheat, puffed | 1C | 54 | 2.3 | 0.2 | 11.8 | 4 | 48 | 0.6 | 1 | 51 | — | 0.08 | 0.03 | 1.2 | — | 2458 |

TABLE 25 (continued)

Calculation of Food Energy Intake. Nutritive Value of Common Foods[a]

| | Amount | Cal | Protein (g) | Fat (g) | Carbohydrate (g) | Calcium (mg) | Phosphorus (mg) | Iron (mg) | Sodium (mg) | Potassium (mg) | Vitamin A (IU) | Thiamin (mg) | Riboflavin (mg) | Niacin (mg) | Ascorbic acid (mg) | USDA Handbook No. 456/ Ref. no. |
|---|---|---|---|---|---|---|---|---|---|---|---|---|---|---|---|---|
| **Pasta, Enriched** | | | | | | | | | | | | | | | | |
| Noodles,[b] macaroni, spaghetti | ½C | 91 | 3.0 | 0.6 | 18.3 | 7 | 41 | 0.7 | 1 | 43 | — | 0.11 | 0.07 | 1.2 | — | 1378c, 1299c, 2159c |
| **Meat, Poultry, Fish** | | | | | | | | | | | | | | | | |
| Beef,[b] Lamb, veal | 1 oz | 79 | 7.4 | 5.2 | — | 3 | 32 | 0.8 | 16 | 74 | 2 | 0.03 | 0.07 | 1.6 | — | 353d, 1185e, 2370e |
| Beef liver | 1 oz | 65 | 7.5 | 3.0 | 1.5 | 3 | 135 | 2.5 | 52 | 108 | 15130 | 0.07 | 1.19 | 4.7 | 7.7 | 1267a |
| Bologna (1 slice, 4½" diameter) | 1 oz | 86 | 3.4 | 7.8 | 0.3 | 2 | 36 | 0.5 | 369 | 65 | — | 0.05 | 0.06 | 0.7 | — | 1982g |
| Chicken, light meat | 1 oz | 50 | 9.4 | 1.0 | — | 3 | 80 | 0.4 | 19 | 123 | 17 | 0.01 | 0.02 | 3.5 | — | 682d |
| Cod,[b] haddock, halibut (broiled) | 1 oz | 48 | 6.9 | 1.8 | 0.5 | 8 | 73 | 0.3 | 40 | 121 | 80 | 0.01 | 0.02 | 1.4 | — | 795d, 1118d, 1104d |
| Ham, cured | 1 oz | 92 | 6.3 | 7.1 | — | 3 | 52 | 0.8 | 227 | 71 | — | 0.15 | 0.06 | 1.1 | — | 1779f |
| Hot dog | 1 | 139 | 5.6 | 12.4 | 0.8 | 3 | 60 | 0.9 | 495 | 99 | — | 0.07 | 0.09 | 1.2 | — | 1994c |
| Pork, fresh | 1 oz | 103 | 6.8 | 8.1 | — | 3 | 73 | 0.9 | 17 | 78 | — | 0.26 | 0.07 | 1.6 | — | 1716e |
| Shrimp | 1 oz | 37 | 7.7 | 0.4 | 0.2 | 37 | 84 | 1.0 | — | 39 | 20 | — | 0.01 | 0.6 | — | 2045c |
| Tuna, canned in oil | 1 oz | 82 | 6.9 | 5.8 | — | 2 | 83 | 0.3 | 227 | 85 | 26 | 0.01 | 0.02 | 2.9 | — | 2323h |
| Egg, large | 1 | 82 | 6.5 | 5.8 | 0.5 | 27 | 103 | 1.2 | 61 | 65 | 590 | 0.05 | 0.15 | — | — | 968b |

53

| Food | Serving | | | | | | | | | | | | | | Ref. |
|---|---|---|---|---|---|---|---|---|---|---|---|---|---|---|---|
| **Cheese** | | | | | | | | | | | | | | | |
| Cheddar, domestic | 1 oz | 113 | 7.1 | 9.1 | 0.6 | 213 | 136 | 0.3 | 198 | 23 | 370 | 0.01 | 0.13 | — | 646p |
| Cottage cheese, small curd, creamed | ½C | 112 | 14.3 | 4.4 | 3.1 | 98 | 160 | 0.3 | 241 | 90 | 180 | 0.03 | 0.26 | 0.1 | 647d |
| Peanut Butter | 2 T | 188 | 8.0 | 16.2 | 6.0 | 18 | 122 | 0.6 | 194 | 200 | — | 0.04 | 0.04 | 4.8 | 1499f |
| **Dried Beans and Peas** | | | | | | | | | | | | | | | |
| Navy beans,^ kidney beans, split peas | ½C | 114 | 7.5 | 0.4 | 20.6 | 32 | 123 | 2.2 | 8 | 342 | 15 | 0.12 | 0.07 | 0.8 | 115b, 161d, 1533a |
| **Fats** | | | | | | | | | | | | | | | |
| Bacon, crisp, slices | 2 | 86 | 3.8 | 7.8 | 0.5 | 2 | 34 | 0.5 | 153 | 35 | — | 0.08 | 0.05 | 0.8 | 126d |
| Cream,^ light, 20% or half & half | 1T | 26 | 0.5 | 2.5 | 0.7 | 16 | 13 | — | 7 | 19 | 100 | — | 0.02 | — | 929b, 928b |
| French or Italian dressing^ | 1T | 75 | 0.1 | 7.6 | 1.9 | 2 | 2 | 0.1 | 267 | 8 | — | — | — | — | 1932b, 1936b |
| Margarine/butter^ | 1t | 34 | — | 3.8 | — | 1 | 1 | — | 46 | 1 | 160 | — | — | — | 1317d, 505d |
| Mayonnaise,^ salad dressing | 1T | 83 | 0.2 | 8.8 | 1.3 | 3 | 4 | 0.1 | 86 | 3 | 35 | — | — | — | 1938b, 1940b |
| Oils | 1T | 120 | — | 13.6 | — | — | — | — | — | — | — | — | — | — | 1401j |
| **Nuts** | | | | | | | | | | | | | | | |
| Unsalted peanuts, pecans, walnuts, almonds | 2T | 103 | 2.9 | 9.6 | 2.9 | 20 | 64 | 0.5 | 19 | 102 | 5 | 0.07 | 0.05 | 1.0 | 1496b, 1536j, 2421e, 81 |
| **Desserts** | | | | | | | | | | | | | | | |
| Brownies, with nuts (1¾" × 1¾" × 7/8") | 1 | 97 | 1.3 | 6.3 | 10.2 | 8 | 30 | 0.4 | 50 | 38 | 40 | 0.04 | 0.02 | 0.1 | 813 |
| Cake, chocolate (3"×3"×2") | 1 piece | 322 | 4.2 | 15.1 | 45.8 | 65 | 121 | 0.8 | 259 | 123 | 130 | 0.02 | 0.09 | 0.2 | 525c |
| Cake, plain (3"×3"×2") | 1 piece | 313 | 3.9 | 12.0 | 48.1 | 55 | 88 | 0.3 | 258 | 68 | 150 | 0.02 | 0.08 | 0.2 | 534b |
| Chocolate pudding | ½C | 193 | 4.1 | 6.1 | 33.4 | 125 | 128 | 0.7 | 73 | 223 | 195 | 0.03 | 0.18 | 0.2 | 1823 |
| Cookies, chocolate chip | 2 | 99 | 1.1 | 4.4 | 14.6 | 8 | 24 | 0.4 | 84 | 28 | 26 | 0.01 | 0.01 | 0.1 | 818b |
| Custard, baked | ½C | 153 | 7.2 | 7.3 | 14.7 | 149 | 155 | 0.6 | 105 | 194 | 465 | 0.06 | 0.25 | 0.2 | 948 |
| Gelatin dessert | ½C | 71 | 1.8 | — | 16.9 | — | — | — | 61 | — | — | — | — | — | 1032b |

TABLE 25 (continued)

Calculation of Food Energy Intake. Nutritive Value of Common Foods[a]

| | Amount | Cal | Protein (g) | Fat (g) | Carbohydrate (g) | Calcium (mg) | Phosphorus (mg) | Iron (mg) | Sodium (mg) | Potassium (mg) | Vitamin A (IU) | Thiamin (mg) | Riboflavin (mg) | Niacin (mg) | Ascorbic acid (mg) | USDA Handbook No. 456/ Ref. no. |
|---|---|---|---|---|---|---|---|---|---|---|---|---|---|---|---|---|
| **Desserts (continued)** | | | | | | | | | | | | | | | | |
| Ice cream, vanilla | 4 fl oz | 129 | 3.0 | 7.0 | 13.9 | 97 | 77 | 0.1 | 42 | 121 | 295 | 0.03 | 0.14 | 0.1 | — | 1139d |
| Pie, apple (9" diameter) | 1/8 pie | 302 | 2.6 | 13.1 | 45.0 | 9 | 26 | 0.4 | 355 | 94 | 40 | 0.02 | 0.02 | 0.5 | 1 | 1566c |
| Sherbert, orange | 4 fl oz | 130 | 0.9 | 1.2 | 29.7 | 16 | 13 | — | 10 | 21 | 60 | 0.01 | 0.03 | — | 2 | 2041b |
| Vanilla wafers | 5 | 93 | 1.1 | 3.2 | 14.9 | 8 | 13 | 0.1 | 51 | 15 | 25 | 0.01 | 0.02 | 0.1 | — | 833b |
| **Sweets** | | | | | | | | | | | | | | | | |
| Milk chocolate | 1 oz | 147 | 2.2 | 9.2 | 16.1 | 65 | 65 | 0.3 | 27 | 109 | 80 | 0.02 | 0.10 | 0.1 | — | 587 |
| Molasses,[b] jams, jelly, maple syrup | 1T | 52 | 0.1 | — | 13.4 | 16 | 3 | 0.8 | 2 | 17 | — | — | 0.01 | — | — | 2050b, 1149e, 2049d |
| Soft drinks | 6 fl oz | 72 | — | — | 18.5 | — | — | — | — | — | — | — | — | — | — | 404a |
| Sugar | 1T | 46 | — | — | 11.9 | — | — | — | — | — | — | — | — | — | — | 2230b |

a  Approximate values. All values have been rounded to the nearest decimal point.

b  Average value for the group of foods listed.

c  Dash indicates a true amount of nutrient that contributes to serving size.

From U.S. Department of Agriculture, Nutritive value of American foods, in *Agriculture Handbook No. 456*, Washington, D.C., Government Printing Office.

body fat is the major energy source. The energy value of body fat is 3500 kcal/lb. Let us assume, then, that the prospective dieter is a woman weighing 250 lb, and that she wants to lose 80 lb. Her daily energy intake to maintain her weight is estimated from her diet record (diet diary) to be 2800 kcal. If the food energy level of her prescribed diet is 1200 kcal, then the deficit which must be supplied by metabolism of body fat is 2800 minus 1200 = 1600 kcal/day. Weight loss, in theory, would then be at a rate of 1600/3500 lb per day, or approximately 0.46 lb per day. At this rate, it would take the woman about 6 months (174 days) to lose the desired amount of weight. This schedule assumes compliance with the diet, a constant rate of weight loss, and maintenance of activity at the prediet level. In order to speed up weight loss either the diet must be more stringent, i.e., food-energy intake per day must be reduced, or a program of regular exercise must be followed. Further, the rate of weight loss is not in fact constant but diminishes as pounds are lost. Therefore, for each 25 lb lost, the energy deficit is 100 kcal or 0.03 lb less. Hence, if the food-energy intake remains constant on the diet, the rate of weight loss will be about 0.03 lb per day less for each 25 lb of weight reduction.[115] In fact, it has been shown that for each 25 lb of excess weight, the body requires 100 kcal for maintenance.

Experience has demonstrated that adherence to a prescribed diet for a period of more than 6 months is problematical. Instead of consuming the prescribed level of food energy day by day, eating behavior is erratic with self-imposed fasts and food escapes.

Dietary compliance is assisted by the following support systems:

1. Group weight reduction programs, such as Weight Watchers or TOPS
2. Contractural arrangements with the dieter. For example, the patient may be offered a job if a certain amount of weight is lost or job tenure is dependent on weight loss.
3. Behavior modification

Behavior modification implies that the patient is taught to change eating habits so as to avoid the temptation of eating at times and in places where intake may be overlooked, such as when he/she is sitting in front of the TV or when preparing a meal. Awareness of times, situations, and moods which promote the urge to snack, to overeat, or to drink alcoholic beverages can be taught by a health professional trained in behavior modification. The patient who is trying to modify his/her own eating behavior is usually told to record when, where, and why he/she eats, as a method of achieving an increased awareness of food-related activities. Recordkeeping is therefore more likely to be carried out by the more educated and highly motivated patient who also can find the time to keep the required accurate food-activity diary. For women this means that the group that has the highest prevalence of obesity, namely those who are educationally disadvantaged, have the least chance to be able to use the record component of behavior modification methods as self-assistance in changing food to reduce the number of situations which trigger eating. Our experience has been that a chance to alter eating behavior is a class privilege denied to the indigent, not only because this group may be less well educated, but also because from the practical standpoint, distinction of an eating place and specific eating times, often required of behavior modification, is difficult for people who have to eat in an overcrowded environment.[116]

Patients who will follow regular exercise programs involving walking, jogging, cycling, calisthenics, swimming, or group sports are likely to be moderately overweight. The sedentary, massively obese person is unlikely to change his/her activity pattern much and indeed, may only be able to achieve a moderate increase in physical activity

TABLE 26

**Effects of Different Types of Exercise on Energy Expenditure**

|  | kcal/min | kJ/min |
|---|---|---|
| In bed asleep or resting | 1.08 | 4.52 |
| Sitting quietly | 1.39 | 5.82 |
| Standing quietly | 1.75 | 7.32 |
| Walking 3 miles/hr (4.9 km/hr) | 3.7 | 15.5 |
| Walking 3 miles/hr (4.9 km/hr) with a 10-kg load | 4.0 | 16.7 |
| Office work (sedentary) | 1.8 | 7.5 |
| Domestic work |  |  |
| Cooking | 2.1 | 8.8 |
| Light cleaning | 3.1 | 13.0 |
| Moderate cleaning (polishing, window cleaning, chopping firewood, etc.) | 4.3 | 18.0 |
| Light industry |  |  |
| Printing | 2.3 | 9.6 |
| Tailoring | 2.9 | 12.1 |
| Shoemaking | 3.0 | 12.6 |
| Recreations |  |  |
| Sedentary | 2.5 | 10.5 |
| Light (billiards, bowls, cricket, golf, sailing, etc.) | 2.5—5.0 | 10.5—21.0 |
| Moderate (canoeing, dancing, horse-riding, swimming, tennis, etc.) | 5.0—7.5 | 21.0—31.5 |
| Heavy (athletics, football, rowing, etc.) | 7.5 + | 31.5 + |

From FAO, World Health Organization, Tech. Rep No. 522, 1973; **Bray, G. A., Ed.,** Obesity in Perspective, National Institutes of Health, Department of Health, Education and Welfare, Publ. No. (NIH)75-708, Bethesda, Md., 1973.

without inducing breathlessness. Effects of different types of exercise on energy expenditure and predictable weight loss are shown in Table 26. Modest increases in energy expenditure are usually a goal of behavior modification techniques.

Patients are sensitized to a greater awareness of the ways in which they spare themselves from physical exertion. Encouragement is given to patients not to delegate tasks within the home, not to drive short distances which could be easily reached by walking, and not to seek work and leisure-time activities which are best performed in the sitting position. Activity patterns can be altered by change of occupation and thus the weight loss that follows may be the direct result of extra energy expenditure or may be due to a combination of factors including the new exercise but also due to a lesser number of work breaks for snacks or greater emphasis by workmates or bosses on the desirability of weight control.

When very obese men and women are recommended to join exercise classes, it is important that referral is to a community group which includes other patients of a similar weight range.

If competition to perform is only with slimmer sisters or brothers, then the chances are that the patient will become embarrassed or discouraged or both, and will soon stop attending the class. The idea that exercise(s) increases appetite is erroneous.

Success in diet-exercise-behavior modification programs means not only that weight is lost but also that the reduced weight is maintained. Patients who report many attempts at weight reduction with subsequent return to the previous state of obesity become discouraged. Long-term weight control is most likely if the patient has a strong reason to lose weight and when the reason does not fade with time. It is indeed the responsibility of a good therapist to reinforce motivation.[117]

Claims that weight can be lost quickly and without prolonged self-discipline are appealing. Two methods of weight reduction which acquire and sometimes retain popularity are fad diets and weight-reducing medications. A summary of the aims, use, misuse, and risks of fad diets and weight-reducing pills follows.

**Fad Diets** — The "success" of fad diets rests on the principles that (1) the diet is low in food energy, (2) the diet induces loss of appetite, (3) the diet produces a state of early satiety, (4) the diet produces dehydration, and (5) the diet is attractive because it contains prestigious foods or foods which have sensory appeal. Rigorous low-carbohydrate meat diets obey this principle. Risks of these diets are that they may induce ketosis, precipitate attacks of gout, or contribute to hypercholesterolemia with attendant risks of atherosclerotic heart disease. Other risks are that if foods such as cereals, fruits, juices, and green and yellow vegetables are excluded because of their carbohydrate content, the diet will be deficient in vitamins A and C, as well as thiamin, riboflavin, and folic acid.[118] The history of fad diets was reviewed by Solomon[119] and has been summarized by Overton and Lukert.[120]

Recent notoriety of the various "liquid protein diets" has arisen because of a number of reports of sudden death in women on this regime. The diet was developed on the theory that it is possible to achieve weight loss with preservation of lean body mass by a protein-sparing modified fast, in which 1.2 to 1.4 g protein in the form of lean beef or egg albumin per kilogram body weight is the protein source. However, in the liquid protein diets, which were developed commercially and advertised, the protein source was a liquid extract of beef hide. Liquid protein diets became known to the easily persuaded obese public through the book *The Last Chance Diet Book.*[121] Deaths have been from intractable ventricular arythmias without a history of previous heart disease. In some cases, potassium deficiency has been present at the time of the acute terminal illness. Loss of lean body mass has been found in patients on this diet, despite claims that the diet offered protection against wasting.[122] Whereas the medical profession has been warned of the dangers of the liquid protein diet, both by reports of fatalities resulting from its use which have appeared in widely read journals and by FDA communications, it is anticipated that sale will be prohibited. However, the health professionals have responsibility to acquaint themselves early with potential dangers of new weight-reducing diets, lest such tragedies recur.

**Weight-reducing drugs** — Prescriptions and over-the-counter drugs intended for appetite control or induction of weight loss have had appeal in the management of obesity for many years. The drugs currently in use for these purposes can be divided into four main groups, which include centrally acting anorectic agents (anorexiants), hormones, diuretics, and bulking agents.

*Anorectic agents,* in common medical use, are unsubstituted or substituted amphetamines. The site of action of the unsubstituted amphetamines such as dextroamphetamine is the hypothalamus and function is mediated by adrenergic mechanism. Substituted amphetamines, such as fenfluramine (Pondimin®, Robbins), possess central activity which is located outside the hypothalamus, and function is mediated by the neurotransmitter, serotonin. Drugs with action similar to amphetamine which are used in weight control include benzphetamine (Didrex®, Upjohn), chlorphentermine (Pre-Sate®, Warner/Chilcott), diethylpropion (Tenuate®, Tepanil®, Merrell-National,

Riker), phenmetrazine (Preludin®), phentermine hydrochloride (Wilpo®) and phentermine resin (Ionamin®, Pennwalt).

Recent studies have suggested that dextroamphetamine and related drugs which have similar pharmacological activity depress appetite, whereas fenfluramine produces early satiety. In obesity, effects of any of these drugs on food intake is greatest when they are first taken. With continued use the dose may have to be increased to produce the initial appetite suppressive effect. Prolonged administration is contraindicated because of the need for escalating drug dosage and the associated risk of side effects. Further decreased appetite may not produce a desired decrease in the consumption of high-energy foods that are particularly tempting.

Side effects include nervousness, restlessness, insomnia, rapid heart rate (tachycardia), and hypertension. Risks of abuse and acute overdosage cannot be overlooked. The earlier claim that fenfluramine is without central side effects has not been substantiated.

Phenylpropanolamine is a mild ephedrine-like drug with some central stimulant properties. Whereas it has been claimed that this drug can suppress appetite and it is widely available as a nonprescription weight-reducing agent, it is dubious that in the recommended dosage it has any suppressive effect on appetite.[123,124]

*Hormones,* especially thyroid preparations, have been used in the treatment of obesity mainly because of a mistaken belief that fat people have a low metabolic rate. Dried thyroid gland pills or thyroid hormone preparations are specific for the treatment of hypothyroidism and should not be used in the treatment of obesity. Risks are that signs of hyperthyroidism may be induced, and that particularly in middle-aged women with existent heart disease, cardiac decompensation and heart failure may result.

Human chorionic gonadotrophin (HCG) preparations, recently heralded as a useful adjunctive antiobesity therapy when combined with a 500-kcal diet has been found to have no intrinsic effects either on appetite, food intake, or body weight. Success with HCG-diet regimes has been due to the diet which, because of its stringency, imposes the risk of nutrient deficiencies and therefore is not appropriate for long-term maintenance.[125]

*Diuretics* are never justified for the treatment of obesity. Loss of extracellular fluid will occur when these drugs are taken. Risks of potassium and sodium depletion must not be overlooked. Over-the-counter diuretic preparations consist of ammonium chloride, which is particularly hazardous to obese alcoholics with liver disease. Diuretics do not have any effect on the amount or distribution of body fat.[126]

*Bulking agents* such as methylcellulose or guar gum, when given before food, may possibly swell inside the stomach and produce early satiety and, therefore, reduction of food intake. In studies which have shown a positive effect of bulking agents on weight-loss induction, high dosages have been necessary. "Slimming foods" are based on use of bulking agents or low-nutrient density mixtures containing dietary fiber sources.[127]

### 3. Surgical Treatment of Obesity

Morbid obesity has been treated by jejuno-ileal bypass. This surgical approach to obesity is effective in inducing remarkable weight loss because the absorptive capacity of the small intestine is reduced. Operative risks are high and late effects of surgery include life-threatening nutritional and metabolic diseases. The most prevalent long-term complications of the jejuno-ileal bypass operation are hepatic dysfunction associated with a fatty liver and osteomalacia. Jejuno-ileal bypass surgery should only be carried out where the surgical group is supported by a nutrition team who can monitor the patient at follow-up. Indications for surgery are morbid obesity which has failed to respond to diet therapy with or without symptoms of the Pickwickian syndrome.[128-130]

In view of the fact that treatment of obesity may not be successful, and certain new methods confer risks to health and life, the aim of health professionals should be to provide information to their patients on obesity prevention.

### 4. Guidelines for Obesity Prevention

Education in relation to obesity prevention is the responsibility of health professionals. In patient and client counseling the following information should be conveyed. General instructions:

1. Avoid overfeeding in early life.
2. Avoid eating when hunger is satisfied.
3. Be aware of what you are eating.
4. Maintain energy expenditure by physical activity at a level to balance energy intake.
5. Keep in mind that weight control is a key component of health maintenance.

Specific instructions to parents:

1. Breast-feed and don't give your baby solid food during the first 3 months of life.
2. Don't bribe or reward children with food.
3. Don't insist that a meal must be finished when your child has indicated that he or she has eaten enough.
4. Introduce your family to active games in early childhood and participate in these games with them.
5. Review the eating habits of your family. Cut down on the size of meals and the number or types of snacks. Are the snacks really necessary?

Specific instructions to patients:

1. Don't eat because you are bored.
2. Avoid social gatherings when you are expected to try all the varied foods.
3. Don't reward or comfort yourself with food.
4. Learn the energy (caloric) content of the foods you eat and the nutrients these foods contain.
5. Try to eat foods with a high nutrient-to-calorie ratio.
6. Read food labels to determine the energy content of these foods.
7. Obtain and read cookbooks giving information on the caloric and nutrient content of the recipes. Avoid recipes of high caloric content.
8. Don't eat while working, or work while eating.
9. Don't snack while watching TV or while enjoying spectator sports.
10. Walk and/or run every day. Jog and bike if the weather is suitable. If not, exercise indoors for at least 30 min a day which may include active housework.
11. Don't ask others to go upstairs for you.
12. Make housework more vigorous, including scrubbing, polishing, bedmaking, etc.
13. Don't stop exercising because you are 30, 40, 50, 60, 70, or pregnant, unless so advised by your physician.
14. Learn and practice health maintenance. Remember that obesity is associated with hypertension, heart disease, gall bladder disease, and joint pain. Keep your weight in the right range for YOUR height, age, and sex.

## B. Nutrition and Cancer

There is evidence that both undernutrition and overnutrition can influence cancer development in different body sites. Chronic iron deficiency can produce epithelial changes in the pharynx and upper part of the esophagus which are precancerous. Iodine deficiency causes goiter, and the cellular overgrowth of the thyroid gland which occurs in this condition, as an attempt at compensation for iodine lack, may undergo malignant change.

However, present concern is for the role of overnutrition in cancer, particularly in industrialized countries such as the U.S. Wynder has presented data showing epidemiological evidence that overnutrition plays a role in the development of cancer of the colon, pancreas, kidney, breast, ovary, endometrium, and prostate.[131] His definition of overnutrition does not include the total energy intake but rather excessive intake of specific nutrients. Breast cancer death rates in different parts of the world are related to the level of fat intake with cancer of the breast being an uncommon cause of death in such low fat-intake countries as the Philippines, Sri Lanka, and Japan, and a commoner cause of death in high fat-intake countries such as the U.S., the U.K., the Netherlands, and New Zealand. In rats, breast tumor incidence is also related to fat intake. It is proposed that perhaps high fat intake influences hormone synthesis and hence tumor development.

Animal studies have also shown that two fecal bile acids have tumor-promoting properties in large bowel cancer. Fecal bile acids arise from degradation of primary bile acids by gut bacteria. The bacteria in the large bowel is influenced by food intake, and the bacteria (clostridia) which degrade primary bile acids to the secondary bile acids — which are cancer-promoting — increase with high meat intake. Cancers associated with obesity are cancer of the endometrium and cancer of the kidney in women.

These kinds of support for association between excessive intake of specific foods or nutrients and cancer are difficult to evaluate. While serious examination of the data is warranted we must also ask ourselves whether necessary fat intake, for example, also measures a higher level of intake of a nonnutrient environmental chemical, which is carcinogenic or which promotes tumor growth.[131]

## C. Fat, Cholesterol, and Coronary Heart Disease

Epidemiological studies have linked a high incidence of coronary heart disease to specific dietary patterns, particularly in affluent populations. Excessive intake of dietary cholesterol, saturated fat, and excess food energy prevail in industrialized countries such as the U.S., where coronary disease is a common cause of death. Risk factors for coronary heart disease include hypercholesterolemia (high blood cholesterol). Hypercholesterolemia is so common in U.S. adults that hospital laboratories report as high as 300 mg/100 m$\ell$ as within normal limits. Hypercholesterolemia in persons who do not have an inborn error in cholesterol or other fat transport appears to result from an imbalance between the intake and excretion of dietary cholesterol. Excessive intakes of saturated fats may be an additional factor.

Mortality from coronary artery disease is related to cigarette smoking, hypertension, and abnormal electrocardiograms showing evidence of heart muscle damage, diabetes mellitus, low physical activity, a behavioral pattern associated with work drive (pattern A behavior), as well as intake of contraceptive steroids. It is to be noted that these factors influence the incidence of heart attacks and may not, except in the case of diabetes, contribute to atherogenesis (degenerative changes in blood vessels).

Although absolute proof is lacking, there is a large body of evidence to suggest that a prudent diet with moderate restriction of food-energy intake and lowered intakes of cholesterol and saturated fat over time, preferably a lifetime, would mitigate against atherogenesis. The risk of heart attacks could be diminished, in susceptible people, by adherence to such a prudent diet as well as by control of dietary factors which influence the development of diabetes. We consider that the major coronary risk factor to coronary prone individuals for myocardial infarction is excessive cigarette smoking. All our public health efforts to change the American diet in the direction of lowering cholesterol and saturated fat intake are unlikely to be successful in controlling mortality from coronary artery disease, unless a comprehensive program is adopted to make cigarettes less available.[132]

Among the desirable changes in the American diet which might reduce the incidence of hypercholesterolemia and hence, coronary artery disease, is an increase in the intake of pectin and perhaps other dietary fiber sources. However, conclusive evidence that such a dietary modification could produce a reduction in serum cholesterol in human subjects is presently lacking.

It is presently proposed in the U.S. dietary goals that a prudent diet with restriction of cholesterol and fat and increase in cereals, vegetables, and fruits as sources of complex carbohydrates and dietary fiber would be food modifications likely to reduce the incidence of coronary artery disease and perhaps also other forms of atherosclerosis. These recommendations reflect the knowledge that people who do eat such diets are less liable to have heart attacks.[132]

## D. Salt and Hypertension

Relationships between sodium intake and hypertension are complex. Surveys of human populations indicate an association between salt intake and incidence of hypertension. In rats, susceptible strains develop hypertension when sodium intake is increased. Within the U.S., blacks who show a salt preference greater than whites from childhood also have a high incidence of hypertension, particularly among males.

However, in the general U.S. population, no direct relationship has been established between salt or sodium intake and blood pressure readings. We assume that if sodium intake has an etiological relationship to hypertension, then, high sodium intake from early life — rather than recent increase in sodium intake — would be more likely to lead to development of hypertension.

Nondietary sources of sodium, including those obtained from medications, are rarely included in estimates of sodium intake. If such extraneous sources of sodium are included, then perhaps a stronger relationship between sodium intake and hypertension may be established. Susceptibility to the development of hypertension with high sodium intake may be dependent on genetic factors.

In essential hypertension, with or without obesity, reduction in sodium intake as well as promotion of sodium excretion, through diuretics, serves to reduce blood pressure. Indeed, in the obese on weight-control reduction diets, reduction in blood pressure may be due to sodium restriction rather than caloric restriction.[133-135]

## E. Diabetes and the Diet

Types of diabetes that are recognized are primary and secondary. Primary diabetes is of two types: (1) insulin-sensitive juvenile diabetes and (2) insulin-resistant maturity onset diabetes. Secondary diabetes occurs when the pancreas is destroyed following perforation of peptic ulcer or with chronic pancreatitis. Juvenile diabetics are commonly underweight; maturity onset diabetics are usually obese. Production of insulin by the pancreas is high in maturity onset diabetes, but there is peripheral resistance to the action of insulin which increases with obesity.

It is generally believed that diabetes mellitus occurs in those who have a genetic predisposition to the disease. Whereas there may be several genetic patterns of inheritance leading to different types of diabetes, most studies suggest that inheritance of the disease is complex. However, even in diabetes-susceptible races such as certain

American Indians and New Zealand Maori, it appears that the disease did not become prevalent until the food supply became plentiful. The association of the disease with a liberal diet may depend on the development of obesity and diabetes as a complication in susceptible persons eating to excess, or there may be specific dietary factors which influence the development of diabetes in these people.[136]

The nature of carbohydrate consumed has been suggested as a contributing factor in the development of diabetes. Studies in Israel of Yemenite immigrants have suggested that increase in the prevalence of diabetes was associated with a change in dietary carbohydrate, from starchy foods such as native breads to high intakes of sugar.[137]

West[138] believes that the increased incidence of diabetes among many American Indian tribes can be linked to increased consumption of sugar, increased incidence of obesity, and decreased exercise. These factors also influence the occurrence of maturity onset diabetes in the general population.

In a review of environmental factors that influence the development of diabetes, Arky comments that there are still many unknowns with respect to the influence of obesity, diet, and infection.[139] He believes that an important future field for investigation is whether there are metabolic differences between obese people who develop diabetes and those who do not. Diabetes has developed in Native American populations who have shifted the type of carbohydrate they consume from starch to sugar, and these same people have shown marked increases in body weights, particularly since they adopted a sedentary way of life. Which of these factors has most influenced the increasing diabetes is not clear.

Four therapeutic goals should be followed in the control of diabetes:

1. Reversal of the diabetic state in maturity onset diabetes by weight reduction
2. Control of blood sugar. In juvenile diabetes, this always requires insulin or diabetes diet
3. Prevention or management of complications
4. Management of diabetes in special physiological or pathophysiological states, e.g., pregnancy or infection

Dietary guidelines which should be followed in juvenile and maturity onset diabetes, respectively, are shown in Table 27.

Long-term control of hyperglycemia both by insulin and diet reduces the risk and the severity of diabetic complications involving the kidney (diabetic nephropathy), eye (diabetic retinopathy), and vascular disorders.[140]

## F. Sugar and Dental Decay

Dental caries consists in progressive localized destruction of dental substance which begins by demineralization of the outer surface of the tooth due to the action of organic acids in the mouth and secondary destruction of dental protein by bacterial action, leading to cavity formation. At a late stage of dental caries, secondary infection develops which may extend from the tooth into adjacent tissues.

It is now known that dental caries results from colonization of unprotected dental surfaces by a specific group of bacteria. These bacteria ferment sugar of dietary origin to yield lactic acid and other organic acid which erodes dental enamel. The most important strain of cariogenic bacteria (*Streptococci mutans*) is responsible for plaque formation. Dental plaque is a film composed of a peculiar sugar polymer derived from sucrose, enveloping the bacteria, which adheres to the tooth surface and from which the carious process extends.

The teeth of the growing child can be made more resistant to dental caries by incor-

TABLE 27

Dietary Guidelines for Treatment of Ambulatory Diabetics

| | Maturity onset obese (not insulin dependent) | Juvenile onset nonobese (insulin dependent) |
|---|---|---|
| Caloric restriction | Yes | No |
| Increase frequency of feedings | Not usually | Yes |
| Consistency in intake of carbohydrate, fat, and protein | Not essential | Very important |
| Meals at the same time every day | Not essential | Most important |
| Extra food for increased exercise | Not required | Usually needed |
| Use food (or sugar) to prevent or treat early signs of hypoglycemia | May be necessary | Most important |
| Give low-carbohydrate diets | Not essential if calories are restricted | Not required |
| Restrict total fat and cholesterol intake | Essential | Important |

From West, K. M., Diet therapy of diabetes: an analysis of failure, *Ann. Intern Med.*, 79, 425, 1973.

poration of fluoride into the dental enamel. Fluoride treatment to achieve this end may be by addition of fluoride at a level of approximately one part per million to low-fluoride drinking water supplies, by administration of fluoride-containing drops or tablets, or by topical application of fluoride to the teeth. However, knowledge of the protective role of fluoride does not detract from the fact that dental decay is closely associated with a high intake of sugar in the diet. Indeed, from epidemiological surveys it is clear that among persons whose teeth have not been protected by fluoride, the incidence and severity of dental caries is most significantly related to the level of sugar consumed. Studies suggest that the eating of chewy sweets, which stick to the teeth, may be a particular risk if dental hygiene is neglected.

In the proposed National Dietary Goals for the U.S., the recommendation that sugar intake should be reduced is based largely on the known association of sugar with dental decay, and the knowledge that if sugar intake is sharply reduced, dental caries will be controlled. We do not advocate change in sugar consumption as an alternative measure to fluoridation, but rather as an additional public health measure to combat the dental caries problem. It should be remembered that after the age of 12 years, fluoride has little protective effect against erosion of dental surfaces because it cannot be incorporated in the enamel. On the other hand, maintenance of a low sugar intake is a continued insurance against dental caries.[141,142]

## G. Coffee, Indigestion, and Antacid Abuse

Complaints of heartburn, flatulence, and indigestion are common in heavy coffee drinkers. Person-to-person variability exists in the amount of coffee which induces these symptoms.[143] In explaining this variability, factors which have to be considered are whether heavy smoking and/or alcohol abuse coexist, whether the person exaggerates or minimizes symptoms, and the level of intake of coffee in relation to body weight.[144] Association of episodic alcohol abuse, heavy smoking, and heavy coffee drinking in people of low socioeconomic status is significant.[145] The symptoms may be due to gastritis caused by alcohol, smoking, or coffee, or by a combination of these factors.

Rosacea, which is a skin condition characterized by capillary dilatation over the nose and central face, is also associated with flushing in response to intake of coffee, alcohol, and hot or spiced foods. Patients with rosacea frequently complain of indigestion but evidence points to association between abuse of coffee or alcohol as a cause of their indigestion rather than a direct relationship between rosacea and gastric symptoms or pathology.[146,147]

Caffeine in coffee (and tea) is a stimulant to gastric secretion.[53] However, since decaffeinated coffee may also potentiate gastric secretion and cause gastric symptoms, it has been inferred that other ingredients of coffee may explain the frequency of indigestion in heavy coffee drinkers.[148]

Modern advertising on TV, radio, and in the newspapers extols the merits of antacids for relief of indigestion. So compelling are the claims of those who sell the antacids that they have a ready market for their wares. A common finding is that heavy coffee drinkers and alcoholics take more antacids. Intake of antacids may be excessive in patients with rosacea. Antacids today usually contain aluminum and/or magnesium hydroxide. When these antacids are taken, dietary phosphate is precipitated as aluminum and magnesium salts which are excreted via the feces. With antacid abuse, this leads to phosphate depletion with muscle weakness and, as a secondary effect of phosphate depletion, adult vitamin D deficiency or osteomalacia may develop.[90]

An awareness of the risks of antacid abuse needs to be widely publicized. Further, it would be greatly preferable if men and women would prevent indigestion by lessening their coffee as well as their alcohol consumption. Rosacea sufferers should be told that progression of the dermatosis is much less likely to occur if they completely cut out both coffee and alcohol.

### H. Constipation — Causes and Management

Constipation is defined as fecal hoarding for a period of more than 3 days with or without subsequent difficult or painful defecation. Causes are (1) delayed transit of feces through the colon, (2) prolonged retention of feces within the rectum, and (3) obstruction.

Colonic movements are nonpropulsive or propulsive. Nonpropulsive movements turn the colon contents over and over to present the feces to the water-absorbing lining of the large bowel. Propulsive movements include the gastrocolic reflex, which is initiated by food or hot beverages that push feces into the rectum and induce a desire to defecate and mass movement which shifts feces along relatively long segments of the large bowel.

Factors delaying transit in the large bowel are excessive fecal dehydration and small size of the fecal mass, which induces less desire to defecate.

Common causes of delayed fecal transit are (1) low-fiber diets, (2) pregnancy, (3) starvation or semistarvation diets, (4) laxative abuse, (5) drugs, e.g., morphine and related narcotic drugs, and (6) chronic intestinal obstruction.

The common cause of chronic intestinal obstruction is cancer of the colon or rectum. Less common causes of delayed fecal transit are late effects of radiation damage to the abdomen and Hirschprung's disease or congenital megacolon, which is an abnormality of the large intestine affecting the nerve supply.

Habitual postponement of defecation with a subsequent inability to pass feces may be due to:

1. Fear of pain due to fissures, tumors, ulcers, inflammed hemorrhoids, or painful defecation in the postsurgical or postpartum period
2. Anxiety or depression
3. Situational factors including unhygienic toilet facilities, lack of privacy, work commitments, or other time constraints
4. Fecal impaction due to excessive drying of fecal mass or intake of large doses of the drug cholestyramine
5. Debility as in the aged, in chronic disease states, or in patients with chronic malnutrition (cachexia)

Management of constipation includes first identification of the cause. It is essential that an M.D. examine the patient to identify or exclude the presence of cancer or other inflammatory and obstructive causes. Medical and surgical causes should be treated. The patient must be counseled on bowel habits including eating of breakfast and/or drinking of a warm beverage at the same time every day so that there is enough leisure and seclusion for defecation in comfort.[149]

The diet should be modified to include cereal fiber, particularly coarse bran which may be incorporated into bread or muffins, eaten as breakfast cereal, or added to other foods such as meat.[150] Other foods or beverages which may help to alleviate constipation are yellow and green vegetables. Prunes or prune juice will stimulate colonic movement because prunes contain a natural laxative.

Use of drugs as laxatives is rarely justified, except in patients who require narcotic analgesics for relief of intractable pain.

## I. Risks of Megavitamin Therapy

Megavitamin therapy has been used by physicians for more than 30 years in the treatment of disease. More recently, self-medication with vitamins has been adopted by the public on the assumption that the regime could prevent and cure disease, or that it could result in super health and avoidance of the effects of alcohol and aging.[151] Diseases for which pharmacologic doses of vitamins have been prescribed include tuberculosis of the skin, which, in the late 1940s, was treated by megadoses of vitamin D, dermatoses such as Darier's disease in which megadoses of vitamin A have been given, and genetic and acquired hyperlipoproteinemias as well as Raynaud's phenomena, in which niacin has been administered in massive doses. In each of these treatment modalities, advantages of primary disease control have been greatly outweighed by risks of vitamin intoxication.

Subgroups of the medical profession have also advocated massive intake of niacin, sometimes combined with vitamin C, for schizophrenia. High-dosage injections of vitamin $B_{12}$ have been given to promote appetite in debilitated states or to treat a variety of chronic diseases for which no specific drug was or is available. Acne is quite commonly treated by vitamin A. Hyperactivity in children has been treated by megavitamin A therapy.[152] In none of these conditions has there been sound scientific evidence that megavitamin therapy can alter the course of the disease.

The public, learning from popular magazines and from radio or TV that vitamin E is necessary to fertility or that this vitamin can keep you young in body and mind, has been persuaded to take large amounts on a daily basis. The news that vitamin C could prevent and cure the common cold also prompted purchase and daily consumption by many people of vitamin C in doses of 1 to 3 g or more.

There are actually only a small number of inborn vitamin-dependent diseases, such as homocystinuria, in which it has been shown that there is a need for very high intakes of specific vitamins in order to control abnormalities of critical enzyme function. Vitamin dependency may exist in certain alcoholics with respect to the B vitamin, thiamin.

Through the experiences of these prescribed or self-prescribed vitamin treatments, we have acquired the knowledge that pharmacological effects of vitamins may be entirely different from physiological effects, and that excessive intakes of vitamins may be dangerous. Dangers of megadosage therapy can be considered under the following classification:

1. Hypervitaminoses may develop which are life-threatening.
2. Megadoses of vitamins may interfere with the bioavailability or desired effect of medications.
3. Mistaken belief that vitamins have a controlling influence in disease prevention or progress may divert patients from seeking and obtaining appropriate treatment.
4. Alcoholics may take massive doses of vitamins in order to protect themselves against cirrhosis when, in fact, only abstinence may be effective in this regard, and then only when discontinuation of alcohol takes place before the advent of irreversible liver damage.

Hypervitaminoses are known to occur with overdose of vitamins A, D, K, and niacin.[153,154]

### 1. Hypervitaminosis A

At risk are toddlers who accidentally swallow vitamin capsules, infants and young children with ichthyoses, patients with Darier's disease, and most adolescents or young adults with severe acne. Signs vary with age being different in the infant than in the older child or adult. In infants, clinical features of hypervitaminosis A are evidence of raised intracranial pressure with vomiting and irregular growth of long bones.

Toxic doses of vitamin A to infants are >20,000 IU per day for 1 to 3 months. Unless recognized so that vitamin A intake can be discontinued promptly, hypervitaminosis A is fatal to infants and young children.

Symptoms of chronic vitamin A toxicity in older children and adults are frontal headache, bone pain particularly in the shins, loss of appetite, alopecia with early loss of frontal hair and hair of the eyebrows, dry skin with desquamation and fissures at the corners of the nose and mouth. Jaundice develops in some cases with associated signs of impaired liver function. In adults the toxic dose of vitamin A is 4000 IU/kg body weight per day, and symptoms develop when patients have been on this dose or greater for a period of about 6 months.

### 2. Carotenemia

Carotenemia, which is a yellow pigmentation of the skin, particularly the skin of the palms, soles, and the center of the face, is due to ingestion of very large amounts of carotene from carrots and/or other dietary sources of β-carotene. Although the yellow color may be considered as a sign of vitamin overload, in fact no other ill effects develop. Carotenemia is, however, frequently seen in young women with anorexia nervosa, who may consume carrots to the exclusion of other foods.

### 3. Hypervitaminosis D

Hypervitaminosis D is characterized by hypercalcemia, impaired renal function proceeding to renal failure, and deposition of calcium in soft tissues. In the U.K., hypervitaminosis D occurred in infants at a time when infant formula was too heavily fortified with vitamin D. Milder cases of vitamin D intoxication may be symptom free. More severe cases may exhibit irritability, particularly in infants, excessive thirst (polydipsia), and passage of very large urine volumes (polyuria).

### 4. Pharmacological and Toxic Effects of Niacin

Niacin, as prescribed in the management of Raynaud's phenomena (blanching of the fingers and toes in response to cold), causes flushing. Flushing may be intense and accompanied by a throbbing headache. Niacinamide (nicotinamide) does not have this vasodilator effect and, although equally effective with niacin in the prevention or treatment of niacin deficiency, does not control Raynaud's phenomena. Actually niacin itself has only been shown to offer partial relief of this condition.

Chronic intake of niacin in doses of more than 1 g per day, as used both in the management of hyperlipoproteinemias and in schizophrenia, causes biochemical changes indicative of impairment of cardiac and hepatic function. A characteristic dermatosis having the features of acanthosis nigricans also develops.

### 5. Other Hypervitaminoses

Signs of intoxication from other vitamin overloads are not as well understood. However, massive intake of vitamin C can cause diarrhea, and there has been speculation and tenuous evidence that if a woman takes megadoses of vitamin C during pregnancy the infant may develop scurvy.

Other specific side effects of pharmacologic doses of vitamins include drug interactions and incompatibilities. If folic acid or vitamin $B_6$ are given in pharmacologic doses to patients with seizure disorders, the risk is that seizure control may be lost due to suppressant effects of these vitamins on the bioavailability of anticonvulsant drugs including diphenylhydantoin and phenobarbital.

Large doses of vitamin $B_6$ (pyridoxine) can reduce or obviate the therapeutic effectiveness of L-dopa and of other drugs, including isoniazid (INH) which are vitamin antagonists. Similarly high doses of vitamin K can reduce the effectiveness of coumarin anticoagulants such as Warfarin.

One report in a human subject and animal study showed that continued massive intake of vitamin E can cause potentiation of anticoagulant effects of Warfarin, leading to excessive hypoprothrombinemia and the risk of hemorrhage. High dosages of water-soluble analogues of vitamin K, given to women before delivery or to newborn infants to prevent vitamin K deficiency, may cause jaundice due to hemolytic anemia and brain damage associated with deposition of bile pigments in the brain (kernicterus).

### J. Mineral Overload

The term mineral overload is used here to include both excessive intake of minerals normally required by the human body and effects of toxic minerals which may contaminate food, water, or other substances consumed.[155] Causative factors in mineral overload are

1. Mineral excess in food, beverage, or water supply
2. Contamination of foods from cooking utensils, clays, and soils
3. Accidental overdosage
4. Hyperabsorption and defective excretion
5. Chronic excessive intake from self-prescribed or prescribed medications
6. Inborn errors of metabolism

Specific examples of mineral overload which occur in the U.S. are iron overload which may be acute or chronic, fluorosis, lead poisoning, arsenical poisoning, and Wilson's disease.

### 1. Acute Iron Toxicity

Acute iron toxicity is second only to aspirin overdosage as a cause of poisoning in young children.[156] Reasons are that mothers have attractive-looking or sugar-coated iron pills which are left within reach of preschool children. Warnings of the danger as well as safety caps to pill containers may have lessened the risk, but further preventive measures are desirable.

The lethal dosage is variable, but death may be caused by 2 g of ferrous sulfate. Vomiting, shock, and coma precede death in fatal cases. Gastrointestinal bleeding and convulsions may occur. The prognosis has been improved by the availability of the iron chelating agent, desferroxamine.

### 2. Chronic Iron Overload

Chronic iron overload may be divided into primary and secondary types. The primary form, designated primary hemochromatosis, is a genetic disease of iron transport in which hyperabsorption of iron can be identified. Clinical manifestations of the disease usually develop in middle-aged patients and a preponderance of males are affected. Clinical features include slate-gray pigmentation of the skin, loss of secondary sex hair, diabetes, and cirrhosis. Deposition of iron occurs in the parenchyma of organs and tissues including the liver, pancreas, pituitary gland, and the skin.

Secondary iron overload or secondary hemochromatosis also called hemosiderosis, or siderosis, implies an acquired etiology, though this separation is incomplete.[157] Certain alcoholics with a history of long-term alcohol abuse may develop hemochromatosis. Etiological factors are ingestion of wines high in iron, the presence of the primary hemochromatosis defect in iron absorption, and impairment in iron utilization due to the toxic effects of alcohol.[158] Alcohol can interfere with the synthesis of pyridoxal phosphate in the body which is necessary to iron utilization for hemoglobin synthesis.

In the South African Bantu, who consume native (Kaffir) beers brewed and stored in iron pots, and who have very high iron intakes, secondary hemachromatosis may be associated with scurvy and osteoporosis. The syndrome has been explained as follows: scurvy occurs both because of low intake of ascorbic acid, and because of excessive catabolism of ascorbic acid in the presence of the retained ferric form of iron. Osteoporosis is secondary to the scurvy, and may be due to a defect in bone collagen synthesis.[159] This syndrome is rare or nonexistent in the U.S. Hemolytic anemias including thalassemia major and repeated blood transfusions can, however, contribute to iron overload.

### 3. Endemic Fluorosis

Endemic fluorosis is a term which describes long-term effects of excessive intake of fluoride from the water supply and from food. The concentration of fluoride intake of water which creates a risk for development of endemic fluorosis has been variously stated as 2 to 4 ppm. In fact, this variability in definition of what constitutes an unsafe level of fluoride in water can be explained in that fluorosis is dependent on total fluoride intake, age when high fluoride intake commenced, and duration of intake. Total intake of fluoride from water varies with the total volume of water drunk per day.

Mild fluorosis with signs of dental mottling is seen regionally within the U.S. In certain other parts of the world, notably in the Punjab in India, endemic fluorosis is severe, producing excessive mineralization of bone and bony deformities, particularly in middle-aged or older men who drink large volumes of water to balance heat-induced sweat losses (5 to 6 *l* per day) with a fluoride content in the order of 16 ppm, as well as eating foods which are high in fluoride. Their excessive body burden of fluoride is created over their lifetime.[160]

### 4. Copper Overload

Copper overload, or chronic copper poisoning, used to be seen among people who cooked food in improperly turned copper cooking pots, or those who ate canned food such as peas, to which copper sulfate was added to give color. This hazard has virtually disappeared since cooking in copper vessels has become unusual, and copper sulfate addition to food was made illegal.

On the other hand, copper overload due to the presence of an inborn error of metabolism, does occur. Wilson's disease or hepatolenticular degeneration is a rare metabolic disease with those afflicted having an inherited inability to incorporate copper into the copper transport protein ceruloplasmin.[161] Clinical signs are those of neurological disturbances and cirrhosis. Treatment is by restriction in copper intake and administration of the copper chelating agent penicillamine.

### 5. Arsenic and Lead Overload

Whereas many minerals are essential nutrients at low levels of intake and toxic when consumed at levels which exceed physiological needs, other minerals, particularly heavy metals, if ingested increase tissue levels and at higher levels of intake cause heavy metal poisoning.

Adulteration of health foods with arsenic and lead has been a source of heavy metal intoxication in recent years. Kelp preparations have been found to contain arsenic at levels that can cause chronic arsenical poisoning.[162]

Bone meal, as supplied in health food stores, may have a high lead content, and therefore carries the risk of causing lead poisoning which has been identified as arising from this cause. It has been pointed out that the FDA is only now beginning to address the question of imposing standards for health foods, including, we assume, the banning of those with a heavy metal content.[163]

## SECTION IV.
## EPIDEMIOLOGY OF MALNUTRITION

### A. Common Nutritional Deficiencies

In the U.S. common nutritional deficiencies of men, women, and children outside of hospitals and institutions tend to be subclinical. Nutritional surveys involving national samples (HANES) or local samples have shown a high prevalence of folacin deficiency in all age groups and iron deficiency in special subgroups including preschool children and women during the reproductive phase of life. However, anemias associated with folacin and iron deficiency are not common as community nutritional problems which interfere with functioning capacity. We are concerned that nutrient depletion which is asymptomatic will progress to overt nutritional deficiency if additional nutrient drain is imposed by physiological stress including pregnancy and lactation, by drugs, by surgery or injury, or by infectious or noninfectious disease.[164-166]

### B. Factors Determining Risk of Nutritional Deficiency

Many of the factors which constitute risk factors for nutritional deficiency are the same as those which act as constraints on securing an adequate diet. These include lack of education, poverty, mental or physical disease and disability, aging, alcoholism, drug addiction, food faddism, institutionalization (in acute care hospitals, extended care facilities, prisons, or children's homes), iatrogenic diets, and child abuse. With certain nutritional deficiencies, particularly deficiencies of trace elements, risk factors may be more strictly environmental. In the U.S., iodine deficiency with endemic goiter has been largely wiped out by the public health measure of iodizing salt.[165] On the other hand, fluoride "deficiency", which makes the teeth vulnerable to the development of dental caries, remains because of public resistance to the fluoridation of water supplies.[167]

Physiological stress, as previously stated, increases nutritional requirements. If these are not met because of the coexistence of one or more factors which mitigate against dietary adequacy, then nutritional deficiency is likely to supervene. Examples of such multifactorial causes of nutritional deficiency are iron deficiency anemia in the pregnant adolescent, whose requirements for iron are high and who at the same time restricts her energy intake and hence her iron intake either in an effort to prevent noticeable weight gain and/or because she is ignorant of her own nutritional needs and/or because she is living in circumstances which limit availability of foods of required nutritional value.[168,169] Similarly, requirements for vitamin D are high during the growth period. These requirements are met through synthesis of vitamin D in the skin under the influence of sunlight (UV light), and by consumption of vitamin D-fortified milk. In children with seizure disorders, who are receiving anticonvulsant drugs, vitamin D requirements are increased due to a drug-induced impairment in utilization of the vitamin. If such a child is housebound and is not given milk, rickets will develop.[170]

Risk factors which condition nutritional deficiency are summarized in Table 28.

### C. Nutritional Deficiencies in Geriatric Patients

Nutritional deficiencies in the elderly are determined by diet, disease, and drugs. Intake of food may be reduced if food-seeking ability is curtailed by depression, senile psychoses, alcoholism, imagined poverty, real poverty, transportation difficulties, institution living, late effects of stroke, chronic arthritis, cardiorespiratory impairment due to congestive heart failure or emphysema, and other physical disabilities which render the person housebound. Anorexia, particularly anorexia associated with cancer, severely depresses food intake. Other drugs such as digitalis glycosides may depress

TABLE 28

**Risk Factors Which Condition Nutritional Deficiency**

Socioeconomic
  Poverty-restricted choice of foods
  Lack of education

Personal
  Food taboos and dislikes
  Faddism
  Veganism

Medical
  Alcoholism
  Geriatric disability
  Acute and chronic physical disease
  Psychiatric disease
  Therapeutic diets
  Drugs

food intake because of associated nausea. Heavy doses of phenothiazine tranquilizers may cause elderly patients to become indifferent to food.

These factors not only decrease the quantity of food obtained and consumed but also may limit dietary quality. The nutritional quality of the diet may also be diminished by poor food choice, consumption of ready-to-cook meals of low nutrient content, by restrictive diets, by lack of motivation to prepare nutritious meals, and by nonavailability in the local area of food programs for the elderly. Masticatory inefficiency, due to lack of teeth or dentures, or due to ill-fitting dentures, has been cited as explanation for consumption of an inadequate diet, but we are doubtful because even if old people with this difficulty have problems in chewing meat, they are well able to eat other foods.

The common consequence of inadequate total energy intake in the elderly is cachexia or wasting. Due to the fact that low total intake of food usually implies a low intake of essential nutrients, particularly vitamins, it is not uncommon to see vitamin deficiencies in malnourished elderly patients. Folacin deficiency is quite prevalent in patients in extended care facilities and it has recently been linked with organic brain syndromes. While psychiatric illness, particularly depression, in the elderly may lead to reduced folacin intake and hence folacin deficiency, folacin deficiency may cause mental abnormalities.

Malabsorption of nutrients is not uncommon in the elderly. Causes include impaired synthesis of gastric intrinsic factor leading to malabsorption of vitamin $B_{12}$ and pernicious anemia, late gluten-sensitive enteropathy, congestive heart failure, late effects of extensive intestinal resections for cancer, and chronic radiation enteritis (late effects of radiation to the abdomen). Drugs are frequent causes of malabsorption in this age group, especially excessive intake of laxatives and antacids, as well as intake of cancer chemotherapeutic drugs such as methotrexate.

Mineral depletion has a high prevalence in geriatric patients. Iron deficiency with anemia results mainly from gastrointestinal blood loss either due to cancer or due to high intake of aspirin. Potassium depletion is most often due to long-term use of diuretics. Zinc and magnesium depletion may result from intake of diuretics or may be due to alcohol abuse. Skeletal calcium depletion occurs in association with osteoporosis and osteomalacia. In osteoporosis, bone loss is due to the aging process, but its development is conditioned by long-term calcium intake and by alcohol consumption.

Osteomalacia or vitamin D deficiency in the elderly is due to inadequate exposure to sunlight, lack of vitamin D-fortified milk, and use of drugs including barbiturates, diphenylhydantoin, and glutethimide, which produce calcium malabsorption. Calcium excretion may also be increased by use of nonthiazide oral diuretics.[171]

## D. Malnutrition in the Hospitalized Patient

Malnutrition inside acute care hospitals was forgotten when endemic vitamin deficiencies were conquered, and rediscovered when nutritional assessment capabilities advanced beyond the naked eye detection of terminal cachexia. Prevalence of malnutrition in hospitals depends on the population served, on the quality of nutritional screening, on total patient care, and on the presence or absence of a nutrition support service. Hospital malnutrition may be acute or chronic.[172] Frequent primary and secondary causes of malnutrition are self-imposed or iatrogenic food restriction, inability to eat or retain food, catabolic losses associated with infections or malignant disease, malabsorption syndromes, intravenous feeding, and cancer, as well as drug- or alcohol-induced nutritional deficiencies.[173]

### 1. Nutrition of the Surgical Patient

Among surgical patients, acute malnutrition is common in the preoperative as well as postoperative period. A patient who is unable to eat preoperatively, either because of an acute surgical condition, or because food intake has been forbidden by the surgeon, is nutritionally at risk, and is likely to develop signs of protein malnutrition after the operation if and when oral feeds are still forbidden.

In the previously well-nourished surgical patient, acute protein malnutrition may not be readily recognized by outward appearance since wasting may not be present or may not be apparent. Nutritional risk in patients with acute protein malnutrition arises because synthesis of vital body proteins by the liver is depressed and because a major defense mechanism of the body, namely cellular immune function, becomes deficient. Depressed cellular immune function makes the patient vulnerable to postoperative infection. Due to reduced production of serum proteins, particularly serum albumin, edema will develop. Wound edema, associated with protein malnutrition, retards healing. Further it has been shown that patients with acute or chronic protein malnutrition are at high risk for side effects of anesthesia and other medications. Reasons for susceptibility to adverse drug reactions include decreased rate and efficiency of drug metabolism in the liver and slowed drug clearance from the body.

Chronic protein-energy malnutrition is most likely to be present preoperatively in surgical patients with diseases of the gastrointestinal tract. In adults, cancer of the mouth, throat, esophagus, stomach, or peptic ulcer, inflammatory bowel disease, or cachexia (wasting) may be advanced at the time when a decision is made for surgical intervention. If the disease process is localized to the mouth or throat, food intake has often been markedly reduced due to pain on eating, but in patients with either upper or lower gastrointestinal disease, nutrient gain may have been diminished over time because of vomiting, malabsorption, or increased protein losses via the intestine.

In the cachetic patient, there is not only loss of subcutaneous fat but also of muscle mass. It is not infrequent to see adult patients with cachexia who on investigation are also found to have acute protein malnutrition. Iron deficiency anemia is common in these surgical patients due to low intake of iron as well as blood loss.

Infants with operable birth defects including cleft palate, strictures and fistulae of the esophagus, and pyloric stenosis fail to thrive because of an inability to suck or because of vomiting.

It is important that both adult and pediatric surgical patients with gastrointestinal disease or with other diseases requiring major surgery receive nutritional assessment on admission to hospital. Accurate nutritional assessment and diagnosis are the prerequisites for nutritional intervention which will significantly reduce postoperative morbidity and mortality. If nutritional assessment prior to surgery has been neglected or is not feasible, then assessment should be early in the postoperative period.

Burn patients are particularly liable to become malnourished. Causes of malnutrition and electrolyte imbalance in the burn patient are diminished food and fluid intake, vomiting, hypermetabolic state due to fever, protein catabolism (breakdown) in response to physical stress, caloric deficit due to heat losses from the burn, and loss of fluid and serum proteins from the denuded areas of skin.[174]

Nutritional assessment of surgical patients should follow a standardized protocol which can rapidly generate answers to the following questions:

1. Has the patient lost weight, and if so, how much weight has been lost during the current illness?
2. Is weight loss associated with loss of subcutaneous fat, muscle mass, or both?
3. What is the cause of weight loss?
4. Does the patient show evidence of acute or chronic protein malnutrition?
5. What is the status of his/her immune function?
6. Is the patient anemic?
7. Does the patient have specific vitamin or mineral deficiencies and imbalances?

Pertinent information to be obtained from the patient and/or the family of the patient includes recent change in food and fluid intake, history of vomiting or diarrhea, and usual body weight.

Physical examination should include systems review, determination of edema, height and weight measurement, as well as measurements of midupper arm circumference and triceps skinfold thickness, from which body fat stores and muscle mass (arm-muscle circumference) can be estimated.

Visceral protein status is determined from serum protein values, which should always include determination of serum albumin.

Nitrogen balance determination will assist the surgeon and clinical nutritionist in determining initial protein status and response to nutritional intervention but requires that protein intake is measured and that 24-hr urine collections can be made.

Cellular immune function is examined by skin testing to mumps and candida antigens as well as to streptokinase-streptodornase extracts, to which the patient would normally show a positive skin reaction (redness and induration at the site of intradermal injection) to these antigens. In protein malnutrition, these skin tests are negative.

A complete blood count with determination of red cell indices (Table 29) may indicate whether anemia is present, and, if present, whether anemia is of the microcytic, hypochromic type indicative of iron deficiency, or a macrocytic anemia indicative of folacin or vitamin $B_{12}$ deficiency. Macrocytic anemias occur with malabsorption and may be present in inflammatory bowel disease. In patients who are dehydrated or bleeding or in shock, hemoglobin and hematocrit values cannot be used to evaluate anemia because of hemoconcentration.

While serum electrolyte estimations give valuable information on the state of hydration and indication of imbalances of sodium and potassium, values cannot be used to determine sodium or potassium requirements.

Due to lack of facilities or expertise in most hospital laboratories to carry out other mineral or vitamin studies, requirements for nutritional support can seldom be estimated from serum levels of micronutrients or blood tests demonstrating vitamin status.

TABLE 29

Normal Hematological Values
(M = Male; F = Female)

| Test | | Normals |
|------|---|---------|
| WBC | M | 7.8 ± 3 |
| × 10³ | F | 7.8 ± 3 |
| RBC | M | 5.4 ± 0.7 |
| × 10⁶ | F | 4.8 ± 0.6 |
| Hgb | M | 16 ± 2 |
| g | F | 14 ± 2 |
| Hct | M | 47 ± 5 |
| % | F | 42 ± 5 |
| MCV | M | 87 ± 7 |
| μ³ | F | 90 ± 9 |
| MCH | M | 29 ± 2 |
| μμg | F | 29 ± 2 |
| MCHC | M | 34 ± 2 |
| % | F | 34 ± 2 |

## 2. Nutritional Problems — the Medical Patient

Malnutrition among patients in the medical ward of hospitals may be primary or secondary. Primary malnutrition is here defined as arising from deficient intake or retention of food. Secondary malnutrition implies nutrient depletion due to malabsorption, increased renal losses, decreased nutrient utilization, or increase in unmet nutritional requirements. Whereas primary malnutrition may be due to dietary constraints associated or unassociated with the presenting disease, secondary malnutrition is necessarily related to disease or drugs, or to a combination of these factors.

Primary undernutrition in adolescent and adult patients in the U.S. is most likely to be seen in anorexia nervosa, in patients with paralyses, deformities, or other chronic diseases which interfere with the normal eating process, in cancer patients, and in drug addicts. In infants, marasmus (semistarvation or starvation) is associated with mismanagement of formula feeding and child abuse. Secondary undernutrition is a feature of patients with malabsorption syndromes, advanced malignant disease, or untreated metabolic diseases including untreated juvenile diabetes and hyperthyroidism (Graves' disease).

Protein malnutrition, due to inadequate intake, may be seen in children with phenylketonuria who are maintained on an excessively low phenylalanine diet, and also in children or adults with chronic renal disease whose protein intake has been severely restricted because of impending uremia. Protein-energy malnutrition of the kwashiorkor type may be seen in previously malnourished young children who develop acute infections such as gastroenteritis. The risk of severe protein-energy malnutrition in young children is greatest where the parent or physician has withheld food, because of a mistaken belief that this would shorten the duration of the infection.[175] Protein malnutrition per se in pediatric and adult medical patients is more likely to be due to excessive protein loss. Massive proteinuria (excess protein loss into the urine) occurs in patients with nephrosis. Protein-losing enteropathy, which is a term used to mean a leakage of plasma proteins into the gastrointestinal tract, can also cause severe protein malnutrition. Generalized exfoliative or bullous (blistering) skin diseases such as psoriatic erythroderma or pemphigus can cause protein malnutrition because of massive protein losses from the skin, either by scaling or by loss of serum protein.[176]

Protein or protein-energy malnutrition in medical patients is commonly multifactorial. Protein malnutrition occurs quite commonly in cancer patients whose intake of protein is inadequate, but who also have other causes for protein deficiency, including losses of plasma protein into serous effusions and protein-losing enteropathy.[177] Reduced albumin synthesis by liver cells as well as protein losses into ascitic fluid may coexist with low protein intake as causes of protein malnutrition in patients with alcoholic cirrhosis.[178]

Hypovitaminoses in hospital patients have been identified in recent years by biochemical studies. Prevalence of vitamin depletion is high in alcoholics and in patients with vascular disease. According to Leevy et al.,[179] approximately 40% of patients with hypovitaminemia have a history of inadequate dietary intake. The most common hypovitaminoses found in the Leevy study were of folacin, thiamin, niacin, and vitamin $B_6$.[179]

Multiple vitamin deficiencies occur in patients with malabsorption syndromes. Since vitamins are absorbed at different proximal and distal sites in the small intestine, the spectrum of vitamin deficiencies which occur in a particular patient and syndrome depends on the portion of the small intestine which is involved by the disease process, or if resection has been previously performed, on the absorptive capacity of the intestinal remnant. In pernicious anemia, vitamin $B_{12}$ deficiency is due to malabsorption of this vitamin, because of an absence of the gastric intrinsic factor. In end-stage renal disease, vitamin D deficiency with osteomalacia or rickets is due to failure of the renal process whereby 1, 25-dihydroxy-cholecalciferol, the active metabolite of vitamin $D_3$, is produced in the kidney. Renal patients on hemodialysis may develop acute deficiency of B vitamins such as folic acid or riboflavin, because these vitamins are lost through the dialysis fluid. In alcoholics, admitted for alcohol-related diseases, vitamin deficiencies whether subclinical or clinical, may be due to inadequate intake, malabsorption, hyperexcretion via the kidneys, or excess gastric loss, as well as failure of storage of vitamins by the liver, abnormal enterohepatic circulation of vitamins, or impaired vitamin utilization by the alcohol-damaged liver, bone marrow, and brain. Patients with diseases associated with higher than normal rates of cell turnover, such as those with hemolytic anemias (e.g., sickle cell anemia), may develop folacin deficiency because the demand for the vitamin for maturation of the red cells in the bone marrow is greater than the supply.

Causative mechanisms producing nutritional deficiencies and particularly vitamin deficiencies have been summarized by Herbert[89] under the following headings: inadequate nutrient ingestion, inadequate nutrient absorption, defects in transport and delivery, inadequate utilization, deficiency due to excessive excretion, and increased requirements that are not met by intake.

Common medical diseases associated with malnutrition, most frequently associated forms of malnutrition, and causal mechanisms are summarized in Table 30.

### 3. Nutritional Support of Hospitalized Patients

In order to meet food energy and nutrient requirements of patients on special diets, defined formula feeds, and total parenteral alimentation, the following criteria should be met:

**Energy intake** — Food energy must provide for the patient's needs as related to age, need to maintain growth, sex, need to achieve or regain usual body weight (except if patient is obese), catabolic losses, and aim to prevent acute metabolic complications.

TABLE 30

**Trauma and Common Disease Associated with Malnutrition**

| Trauma/disease | Cause of malnutrition | Nutrients lost |
|---|---|---|
| Mechanical injury or surgery | Food intake↓<br>Catabolic response† | Nitrogen[a]<br>Potassium<br>Zinc |
| Burns | Food intake↓<br>Catabolic response<br>Serum loss<br>Hypermetabolism | Nitrogen<br>Sodium<br>Potassium<br>Zinc |
| Infections | Anorexia<br>Toxemia<br>↓ Fever<br>Catabolic state<br>Negative energy balance | Nitrogen<br>Potassium<br>Zinc |
| Diseases of the intestine | | |
| Cancer esophagus or stomach | Food intake↓<br>Vomiting | All |
| Chronic disorders of the small intestine, e.g., celiac disease | Failure to absorb nutrients | Fat<br>Fat and water-soluble vitamins<br>Calcium |
| Inflammatory bowel disease, e.g., regional enteritis — Crohn's disease | Food intake↓<br>Blood and protein loss via the gut<br>Fistulae, medication | Iron<br>Nitrogen |
| Psychological disease | Food intake↓ | No abnormal losses |
| Anorexia nervosa | Vomiting | |

[a] Leads to protein malnutrition.

**Fluid** — Sufficient fluid must be supplied to correct hydration and provide for the excretion of nitrogenous end products of endogenous compounds and other end products including drugs.

**Protein (nitrogen)** — Protein or nitrogen requirements depend on age, sex, ideal body weight, degree and type of protein malnutrition, continued excessive protein losses, liver and kidney functions, as well as the selected means of nutritional therapy. In general, the major aims are to prevent wasting and to maintain or return the patient to nitrogen balance.

**Vitamins and minerals** — Amounts to be supplied should be at levels to meet or restore a metabolic function. Correction of vitamin deficiencies and mineral imbalances is essential.

Nutritional support of hospital patients may be from diets composed of normal foods, from chemically defined formula, given by mouth or by tube feeding, or by intravenous nutrition. Chemically defined formulas containing amino acids, modified fat sources, and sugars as well as vitamins and minerals are used most frequently in enteral feeding of patients in whom digestion or digestion and/or absorption of nutrients is grossly impaired. Whereas the formulas may be administered via a feeding

TABLE 31

**Indications for Specific Types of Nutritional Support**

**Modified Diets**
Diabetes — energy restriction or redistribution
Heart disease
Congestive heart failure/hypertension, low sodium
Coronary heart disease — low cholesterol, low fat
Gastrointestinal disease
Peptic ulcer — meal frequency†
Celiac disease — gluten free
Inborn errors of metabolism
PKU — phenylalanine restriction
Galactosemia — lactose containing food excluded
**Chemically Defined Diets (Oral or by Tube Feeding)**
Preoperative bowel preparation
Postoperative management of colonic and rectal surgery
Inflammatory bowel disease
Malabsorption and short gut syndromes
Pancreatitis
Hypermetabolic states
**Total Parenteral Nutrition (TPN)**
Diseases or injuries associated with loss of alimentary function
Hypermetabolic states such as severe burns
Inflammatory bowel disease with fistulas
Any disease causing requirement for prolonged intravenous feeding

tube or by mouth, the feeding tube is needed when continuous or nightly nutrient intake is desirable as well as in patients who have difficulty in swallowing or retaining food by mouth.

When enteral feeding (mouth or tube) is inadequate to permit nutritional support or rehabilitation, then parenteral alimentation may be instituted. Parenteral or intravenous nutrition may be via a peripheral vein or through a central venous catheter. When obstruction or surgery of the gastrointestinal tract are temporary contraindications to oral feeding, short-term intravenous feeding by peripheral vein is usually the method of choice. Aims of intravenous nutrition via a peripheral vein are commonly "protein sparing" with the prediction that a return to oral feeding will be at an early and foreseeable date. Total parenteral alimentation via a central venous catheter (TPN) is indicated in severe protein-calorie malnutrition, in hypermetabolic states of long duration, when intestinal absorptive function cannot be restored due to resection or radiation enteritis, and in the infant under conditions of intractable diarrhea. Indications for specific types of nutritional therapy are listed in Table 31.

It has been pointed out by Barlow[180] that in patients whose nutritional requirements are very large, as for example, burn patients, it may be necessary to use parenteral alimentation via a central vein as well as feeding by the oral route in order to prevent nutritional depletion.[180]

Advantages to the patient of defined formula diets or parenteral alimentation have been proven, both with respect to recovery from surgical or medical illness, and with respect to lowering of mortality from medical and surgical causes associated with malnutrition.

A particular indication for total parenteral alimentation is the patient with regional enteritis who has the complication of chronic discharging fistulae (false intestinal openings) on the abdomen which have failed to heal. When complete bowel rest can be afforded as with TPN, the fistulas will close.

Nurses, physicians, and pharmacists responsible for the care of patients receiving defined formula diets or parenteral alimentation should be aware of possible side effects. Barlow[180] warns that use of packaged formulas for enteral and tube feeding requires monitoring of patients for electrolyte and acid-base imbalance, and also to determine intestinal tolerance of the osmotic load offered by the given formulation. Complications of total parenteral alimentation can be summarized under the following headings:

1. Catheter related
    a. Perforation or obstruction of a major blood vessel
    b. Sepsis (local or systemic)
2. Metabolic
    a. Hyperglycemia, due to high concentration of glucose infusion fluid
    b. Hypoglycemia, usually when infusion is stopped
    c. Increased blood urea nitrogen, because of excessive nitrogen load
    d. Hyperammonemia (in alcoholic patients)
    e. Electrolyte imbalance including hyponatremia (sodium depletion), hypernatremia (sodium excess), hypophosphatemia, and hyperchloremic acidosis
3. Allergic reactions
    a. Allergic reactions may occur due to hypersensitivity to thiamin in the infusion fluid
4. Acute deficiency state
    a. Essential fatty acid deficiency
    b. Zinc deficiency
    c. Folacin deficiency (when ethanol is used as an energy source)

Review articles and chapters on indications for defined formula diets, tube feeding, and total parenteral alimentation include those of Taylor,[181] Glotzer et al.,[182] MacFadyen et al.,[183] and Shenkin and Wretlind.[184]

## E. Nutritional Deficiency in People on Special Diets
### 1. Vegan Diets
Vegan diets are diets which do not contain sources of animal protein. Such diets are lacking in vitamin $B_{12}$, heme iron, and calcium. Riboflavin intake from vegan diets may be low because no milk or milk products are consumed. In vegan children, food energy intakes may be lower than in children on nonvegan diets because of difficulty in eating bulky foods.

The relative infrequency of reported vitamin $B_{12}$ deficiency with megaloblastic anemia and neurological complications has been attributed to various causes. More plausible explanations are that vitamin $B_{12}$ deficiency does not develop for 2 to 3 years after cessation of intake of the vitamin, and that a vegan diet may contain animal protein contaminants. However, if for sociocultural or religious reasons a person is committed to eating a vegan diet, then a dietary supplement of vitamin $B_{12}$ is recommended, and he/she should seek advice about their nutrient needs, and about the requirements of their vegan children. The nonnutritionist health professional, who may be asked to give advice, should obtain an expert opinion to help in giving guidance. It is particularly important that careful nutritional advice be given to pregnant women who insist on practicing a vegan regime during pregnancy and lactation because the risk of iron and calcium deficiency is greatest in these periods of physiological stress.[185-187]

## 2. Elimination Diets

Allergists, pediatricians, and dermatologists may prescribe elimination diets in order to find out whether health problems such as infantile eczema or urticaria (hives) are due to food allergy. These diets only contain foods which are known to be unlikely to cause food allergy. The common elimination diet includes lamb, beef, chicken, turkey, peas, rice, applesauce, apple juice, and pears. The diet is nutritionally unsound and should only be employed for a period of 7 to 10 days as a diagnostic measure. The assumption is that if the health problem is due to food allergy, then the signs will disappear when the diet is being given. If the health problem persists, then the diagnosis of food allergy is incorrect. In cases where the signs of the health problem do disappear with the diet, then new foods should be gradually introduced, until the time when the "allergic food" is given and signs reappear. If the total period of major food restriction exceeds 7 days, then the child or adult should receive a vitamin-mineral supplement to insure intake of needed micronutrients. It is important that the vitamin supplement is checked for allergenicity and a preparation should be advised which does not contain the yellow food color, tartrazine, which is a major cause of hives.[188]

## 3. Ulcer Diets

Rigid Sippy diets, which are milk based, are no longer justified in the treatment of peptic ulcer. These diets do not improve the outcome of peptic ulcer, and if maintained, confer the risk that the patient will develop scurvy since the rich vitamin C source is included. If a peptic ulcer patient has been advised to limit the diet to milk or milk and bread and butter with or without fish or chicken, then it should be suggested that a vitamin C supplement be taken each day.[189]

## 4. Diarrhea Diets

Dietary and hence, nutritional, deficiencies may develop because of poor advice to parents or poorly understood instructions to parents for the management of acute diarrheal illnesses occurring in their infants or young children. Food may be entirely proscribed and the mother is told only to give cola drinks or ginger ale. With this regime, protein-energy malnutrition will soon occur. The mother may also not understand the physician's directions about diet in diarrheal illness. An infant seen in the State of New York with kwashiorkor (severe protein-energy malnutrition) had been fed for 3 weeks on Kool-Aid® because the mother understood the doctor to say that this should be the sole intake for a child during a bout of gastroenteritis. It is to be noted that previously the mother had had several of her children placed in foster homes because of parental neglect (unpublished).

## 5. Lactose-Intolerance and Milk-Free Diets

When large volumes of milk are drunk by those who are lactose intolerant, abdominal discomfort and gassy or watery diarrhea will follow. Blacks are more often lactose intolerant than whites.[190,191] In children and adults who are lactose intolerant, care should be exercised in advising elimination of milk from the diet. By excluding milk, the dietary intakes of calcium, vitamins A, D, and riboflavin may be seriously reduced, especially unless other food items containing rich sources of these nutrients are taken. Calcium deficiency with or without added vitamin D deficiency can cause rickets in children.[192]

Further lactose intolerance is usually partial and a limited amount of milk is commonly taken without producing symptoms in lactose-intolerant persons.

## F. Malnutrition in Newborn Infants

### 1. Causes of Malnutrition and Growth Failure in Newborn Infants

Low birth weight infants can be divided into those who are premature and those which have experienced intrauterine growth retardation but are born at term (dysmature infants).[193] Common causes of intrauterine growth retardation are drugs ingested by the mother during pregnancy, maternal infection, maternal alcohol abuse, maternal smoking, toxemia of pregnancy, undernutrition of the mother during pregnancy, and chromosomal, structural, or metabolic defects of the fetus.

Early dysmaturity occurs when the development of the fetus is impaired in the first or early in the second trimester of pregnancy when organs and tissues are being formed. There is a high risk of early dysmaturity when the mother takes medications during the early part of pregnancy which are known to interfere with fetal development. Drugs known to have this effect are cancer chemotherapeutic agents and anticoagulants. Other drugs which are highly suspected as causes of early fetal dysmaturity are anticonvulsant drugs such as diphenylhydantoin, and oral antidiabetic agents. Aspirin and related drugs, such as salicylamide and acetaminophen present in over-the-counter analgesics, produce growth retardation in pregnant laboratory animals and may do so in women. Malformations as well as growth retardation are associated with early dysmaturity.[90,194]

Late fetal dysmaturity occurs when the growth and nutrition of the fetus is impaired during the latter half of pregnancy, after organs and tissues are developed, and during the time of fetal maturation. Undernutrition of the mother is a major cause of late fetal dysmaturity. At birth, infants with late fetal dysmaturity are grossly underweight but do not show malformations.

Maternal smoking is a very important cause of infants being underweight at birth, whether or not the mother is also undernourished.[195] Alcohol abuse by the mother during pregnancy can cause both early and late fetal dysmaturity. Infants of actively drinking alcoholic mothers exhibit the fetal alcohol syndrome. They are small-for-dates at birth, show distinctive minor malformations of the face, may have cardiac defects, and fail to thrive in early life.

High-risk pregnancy factors, that is, factors which have been shown in practice to contribute significantly to poor pregnancy outcome and birth of underweight infants, may be socioeconomic, demographic, medical (related to illness in the mother), or related to disorders of pregnancy. Low birth weight is usually multifactorial. Contributory factors are commonly low socioeconomic and educational status, physiological immaturity, smoking, alcoholism, drug addiction, intake of medications, inadequate diet, and lack of antenatal care. It should be noted that whereas short women and small women are more likely to have small babies, the relationship between shortness of the mother and low birth weight of the infants is strongest for low-income women. Women who have not reached their genetic potential for height are likely to have suffered malnutrition in early life. Hence, effects of maternal height on the infant's birth weight are explained by recurring adverse social circumstances which preclude adequate nutrition of the mother and her developing infant during pregnancy.

Girls who have more than one pregnancy during adolescence are at high risk and if the first of the infants has been underweight, it is more likely that the girl will have a second underweight baby. Obstetrical risks for teenagers include premature births, low birth weight infants, higher neonatal mortality, iron deficiency anemia of the mother, and, less frequently, prolonged or difficult labor due to feto-pelvic disproportion (the infant may be large relative to the mother's pelvic size). Obstetrical performance in young women and in adolescents is greatly improved by good antenatal care. Self-prescription, or MD prescription, of weight-reducing diets in pregnancy imposes the risk of inadequate maternal weight gain and birth of an underweight infant.[196,197]

Nutritional anemia is common in pregnancy. Iron deficiency anemia is due to low intake of iron coupled with blood loss at parturition. Folacin deficiency with anemia is caused by low intake of the vitamin folic acid and the high folacin requirements of the fetus which impose a strain on the maternal supply of the vitamin. Coexistent nutrient deficits of iron and folacin tend to occur in girls under 19 years who have had several pregnancies.[198]

## 2. Nutritional Hazards of Low Birth Weight Infants

Nutritional hazards and disorders of premature and dysmature infants that are of low birth weight can be summarized as follows:

1. Feeding difficulties
    a. Inability to suck
    b. Regurgitation or vomiting of feeds
    c. Inhalation of feed
2. Metabolic disorders
    a. Impaired thermoregulation
    b. Neonatal hypoglycemia
    c. Neonatal hypocalcemia
    d. Protein intolerance
3. Nutritional deficiencies
    a. Hemorrhagic disease of the newborn (vitamin K deficiency)
    b. Folacin deficiency
    c. Vitamin E deficiency
    d. Iron deficiency
    e. Essential fatty acid deficiency

Feeding difficulties are greatest in infants with the respiratory distress syndrome, with congenital malformations such as cleft palate, with congenital heart disease, or with inborn errors of metabolism such as galactosemia.

Impaired heat regulation is due to diminished fat stores. Neonatal hypoglycemia usually arises because of low carbohydrate (glycogen) stores in the liver and failure of the premature or dysmature infant to maintain blood sugar levels in the fasting state. Neonatal hypoglycemia also occurs in the infants of diabetic mothers who, however, usually are large at birth. Signs of neonatal hypoglycemia are high-pitched cry, tremors, convulsions, weak and irregular respirations, failure to suck, muscle weakness, and coma. Unless promptly treated, neonatal hypoglycemia — by reducing the glucose supply to the infant's brain — causes permanent brain damage. Late effects of untreated neonatal hypoglycemia are seizure disorders, cerebral palsy, and mental retardation.[199]

Neonatal hypocalcemia, which is hormonal in origin, may cause tetany leading to convulsions. Protein intolerance arises when premature infants are given feeds with a very high protein content such that the immature kidney cannot excrete the nitrogen wastes. Vitamin deficiencies such as folacin deficiency arise because of inadequate transference of nutrients from the mother to the fetus during pregnancy.

Vitamin E deficiency may occur in premature infants on formula feeds high in polyunsaturated fats. In these infants, a contributory factor in vitamin E deficiency is low body stores of the vitamin. Vitamin E deficiency in premature infants causes hemolysis (bursting of red blood cells), leading to anemia. If premature infants are given iron, hemolysis induced by vitamin E deficiency is worsened.[200]

Vitamin K deficiency occurs in premature infants as well as in full-term infants if the mother has not received vitamin K before delivery, if the mother has been receiving anticoagulant (antivitamin K) drugs of the coumarin group during pregnancy, and also if the infant is not given vitamin K by injection during the first few days of life. Vitamin K-deficient infants bleed at birth or in the early days after birth. However, administration of high dosages of water-soluble analogues of vitamin K to a woman in labor or to an infant soon after birth is dangerous and may result in hemolytic anemia, jaundice, kernicterus (jaundice of nuclei in the brain), and death.[201]

Iron deficiency is more common in infants of low birth weight because the iron content of the body is reduced. Nutritional anemia of low birth weight infants may not only be due to iron deficiency, but also to folacin, vitamin E, or vitamin K deficiency.[202]

Essential fatty acid deficiency may arise in premature or dysmature infants who are fed intravenously without addition of these lipids. Signs are scaliness and redness of the skin, emaciation, and growth failure.[73]

### 3. Nutritional Management of Premature and Dysmature Infants

The most urgent nutritional requirement in the management of premature and dysmature infants is prevention or treatment of neonatal hypoglycemia. Treatment is by intravenous glucose, which should be continued until blood glucose values have stabilized within normal limits. Expressed breast milk feedings are the food of choice. If formula feeding is employed, the mixture must be calorically concentrated, because very small infants with stomachs of very limited capacity will not take large volumes of fluid. Rapid growth may be promoted by feeds providing 100 kcal/100 m$l$, but if such feeds are given, additional water is required to maintain hydration.

Feeds may be given orally, but if the infant's ability to suck is in question, then it is preferable to administer the breast milk or formula by stomach tube.

Vitamin and iron supplements should be given at a level to supply the daily needs of low birth weight infants.[203]

### G. Malnutrition in Alcoholics

Alcoholics may be thin or fat. In different socioeconomic groups, the prevalence of fatness and thinness among alcoholics tends to parallel the distribution of adiposity found in that subgroup of the population. Thus, in women of low socioeconomic group or low socioeconomic stratum (SES) of origin, obesity is common, while in men of the same SES group, thinness is likely to be found. On the other hand, in alcoholics of middle or upper socioeconomic groups, or from such groups of origin, the women tend to be slim and the men tend to be obese. Late alcoholics depart from these generalizations on body size in that they may be grossly underweight with loss of subcutaneous fat and muscle wasting.

Weight loss in alcoholics occurs (1) acutely during an alcoholic spree when little or no food is taken, (2) during alcohol withdrawal, when nausea, vomiting, lack of appetite, or mental disturbances severely reduce all intake of nourishment, and (3) when alcohol-related diseases or disorders are present that decrease appetite, reduce nutrient absorption, increase nutrient losses, and impair nutrient utilization.

Weight gain in alcoholics occurs under the following conditions:

1. When alcohol intake is in addition to a food energy consumption equal to or greater than daily requirements
2. During nutritional rehabilitation, following voluntary or enforced abstinence
3. With development of ascites in patients with alcoholic cirrhosis[90]

TABLE 32

**Factors Leading to Malnutrition in Alcoholics**

1. Depressed food intake
2. Poor choice of foods
3. Maldigestion and malabsorption
4. Hyperexcretion of nutrients
5. Malutilization of nutrients
6. Increased nutrient requirements

TABLE 33

**Common Types of Malnutrition in Alcoholics**

1. Protein energy malnutrition
2. Nutritional anemias (folacin deficiency)
3. Nutritional bone disease
4. Thiamin deficiency (Wernicke-Korsakoff psychosis)

The primary effect of alcohol on the tissues is toxic. Major target organs damaged by alcohol are the liver, pancreas, gastrointestinal tract, bone marrow, heart muscle, and brain.[207-210]

Reversible malabsorption occurs in actively drinking alcoholics due to the acute toxic effects of the alcohol on the mucosa of the gastrointestinal tract. Under these circumstances, absorption of B vitamins, particularly thiamin and folacin, is markedly reduced. Maldigestion and malabsorption occur in alcoholic pancreatitis. Alcoholics with liver disease including hepatitis and cirrhosis have a decreased capacity to store or utilize nutrients, particularly water- and fat-soluble vitamins. Chronic alcoholics become vitamin dependent, which means that for maintenance of brain or bone marrow function, they require higher levels of intake of specific vitamins than nonalcoholics. Prevention or treatment of the Wernicke-Korsakoff syndrome requires that pharmacologic doses of thiamin be administered. Nutritional anemias are not uncommon in chronic alcoholics, usually with combined folacin and vitamin $B_6$ deficiency. Nutritional diseases which are more common in alcoholics than nonalcoholics are iron overload (hemachromatosis), osteoporosis, and zinc deficiency. Zinc deficiency occurs not only because of low intake, but also because urinary losses are increased.

Factors leading to malnutrition in alcoholism are summarized in Table 32. Table 33 classifies common types of nutrient depletion or deficiency occurring in alcoholics.[90,204-206]

Nutritional rehabilitation of alcoholics requires discontinuation of drinking. Liver damage in alcoholics cannot be prevented by provision of a nutritious diet, nor can cirrhosis or other chronic tissue damage due to alcohol be reversed by dietary therapy.

# SECTION V
## DRUG-NUTRIENT INTERACTIONS AND INCOMPATIBILITIES

## A. Food-Related Drug Reactions
### 1. Food Allergies

Food allergies can be to food proteins, to other large molecular weight components of food, or to small molecular weight components which may be naturally present in a food or used as chemical food additives. Nonnutrient natural components of food as well as food additives may be similar to or chemically identical with drugs. Further, flavoring or coloring agents used in foods may be the same that are used for similar purposes in medications. Allergic reactions may develop when people have been sensitized to these small molecular weight substances, either through food or drug ingestion.[211]

Common small molecular weight allergens occurring in foods, beverages, and drugs are ethylenediamine tetraacetic acid (EDTA), salicylates, tartrazine, sodium benzoate, erythrosine, quinine, iodine, penicillin, and menthol. EDTA is used as a stabilizer or to protect flavor in mayonnaise and other salad dressings. Salicylate occurs naturally in certain fruits such as plums. Methyl salicylate is also used extensively as a flavoring agent under the name oil of wintergreen or teaberry. Cereal breakfast foods containing salicylate include Breakfast Squares®, Food Sticks, apple, blueberry, and cherry turn-over pastries, and refrigerated cinnamon-sugar cookies. A number of types of candies contain methyl salicylate including wintergreen, gums, mints, and other flavored candies, jellybeans, and candied "bird eggs". Beverages containing salicylate include root beers (Hires®, Nehi®, Dad's®, and Draft®) and Dr. Pepper®. People who are allergic to aspirin (acetyl salicylic acid) or sodium salicylate may also exhibit identical reactions when foods and beverages containing salicylates are consumed. The most common type of reaction is urticaria, or hives.

Tartrazine is present in many breakfast cereals, in low-protein pasta products, in commercial baked goods, in numerous types of candy including the black licorice sandwich, candy corn, lemon drops, Jujubes® Fruit Drops, halloween candy, and various other yellow and green candies. Tartrazine is also contained in fruit crushes of various flavors, Coca Cola®, and Shasta® diet drinks.[212] In spite of current recommendations for the elimination of tartrazine as an additive to therapeutic drugs, the dye is still present in many drug formulations including capsules — such as antibiotic capsules — and vitamin preparations. As with salicylate, intake of tartrazine by sensitized people leads to the development of urticaria and chronic urticaria may be attributed to this cause. Salicylates, tartrazine, and sodium benzoate may also cause asthma and purpura.[213] The severity of reaction is dependent upon the degree of hypersensitivity and the ingested dose of the foreign compound. Another food-coloring agent, erythrosine, is a photosensitizer.[214] Quinine is another potent cause of allergic purpura in the form of quinine sulfate as a drug, but is also present in tonic water, in Campari, and as a flavoring agent in certain gums and candies. Dermatitis, usually of the face, follows ingestion of foods containing EDTA, if the consumer has already developed sensitivity to this compound by previous contact with industrial chemicals, liquid soaps, disinfectants, detergents, deodorizers or cosmetics containing this compound.[215,216]

Present-day milk and dairy products seldom contain appreciable quantities of penicillin, because of legal constraints on the sale of milk from cows under treatment with this antibiotic for cow mastitis. Nevertheless, it would still be possible for someone with a penicillin allergy to obtain their milk supply from an individual farmer who may use penicillin in treating his dairy cattle.[217] Raw cashew nuts, as sold in health food stores or used in vegetarian restaurants, may contain oleo resin, the sensitizer

TABLE 34

Allergic Reactions to Common Food Chemicals

| Chemical additive | Food source | Reaction |
|---|---|---|
| Salicylate | Sodas | Urticaria (hives) |
| Tartrazine | Candy | |
| Menthol | Breakfast cereal | |
| | Candy | |
| | Gum | |
| EDTA | Mayonnaise | Dermatitis |
| Essential oils | Alcoholic beverages | |
| Balsam of Peru | Chocolates | |
| Erythrosine | Maraschino cherries | Photosensitivity |
| | Fruit cocktail | |
| | Breakfast cereals | |
| | Cherry cake mix | |
| Sodium benzoate | Ketchup | Purpura |
| Salicylate | See above | Asthma |
| Quinine | Tonic water | Purpura |
| | Bitter lemon | |
| | Campari | |

Patients react similarly to aspirin, another salicylate.

similar to the sensitizer found in poison ivy plants. Dermatitis will develop if these nuts are handled or eaten by people who are sensitive to poison ivy and similar plants. A summary of food chemicals which commonly produce allergic reactions and the characteristics of these reactions are shown in Table 34.

### 2. Hypertensive Crises and Monamine Oxidase Inhibitors

Monamine oxidase inhibitor (MAOI) drugs all block oxidative deamination of endogenous and exogenous amines including tyramine and dopamine, which are present in certain foods and wines. After it was shown in 1952 that a hydrazide drug, iproniazid, had mood-elevating properties and that this drug functioned as a monamine oxidase inhibitor, other structurally dissimilar drugs sharing the MAOI property were synthesized and used for their effects in relieving neurotic and psychotic depression. These drugs form complexes with monamine oxidase and inhibit its metabolic degradation.

Since 1963, hypertensive crises have been recognized as the most serious side effect of monamine oxidase inhibitor drugs.[218] These crises follow intake of certain foods including cheese, yeast, or yeast extracts, chickens' livers, pickled herring, broad beans, wines, and beer by patients receiving MAOI drugs.[219,220] Symptoms appear 30 to 60 min after the offending food or beverage is consumed under the circumstance of food intake. Complaints are of headache, palpitations, nausea, and vomiting. Transient but often severe hypertensive attacks occur and cerebral hemorrhage is reported as the cause of death in fatal attacks.

Attacks vary with the MAOI drug, dosage, and the tyramine intake from cheeses, wines, or from other tyramine-containing foods known to cause these reactions.[218,221]

In broad beans the amine apparently responsible for inducing hypertensive attacks in MAOI drug users is dopamine.[53] Tyramine reactions are more common, particularly with consumption of cheese or Chianti wine by men and women receiving these drugs. Tyramine metabolism in the liver is inhibited by MAOI drugs. Nonmetabolized tyramine releases norepinephrine from nerve endings, where it is believed that this catecholamine is present in higher than normal concentrations under the circumstance of MAOI drug use.

Prescription of MAOI drugs, as antidepressants has been limited since 1963 when the risk of hypertensive crises was first recognized. Drugs in this group presently available in the U.S. are isocarboxazide, nialamide, phenelzine, and tranylcypromine. Tyrer et al. have advocated use of MAOI drugs in the treatment of phobic anxiety in combination with the tricyclic antidepressant, amitriptyline, or trimipramine, in "refractory" depressive illnesses.[222]

Procarbazine, a drug of choice in the treatment of Hodgkin's disease, is a weak MAOI drug. Hypertension has been described in patients eating tyramine-containing foods while on procarbazine therapy.[223]

The tyramine content of foods and beverages is shown in Table 35. Patients who are receiving MAOI drugs should refrain from eating or imbibing these food items.

### 3. Acetaldehyde Reactions

Tetraethylthiuram disulfide (Disulfiram/Antabuse®) is used in the aversive treatment of alcoholism. If alcoholic beverages are drunk when someone is receiving this drug, disagreeable symptoms occur within 5 to 15 min. Flushing of the face, neck, and upper chest is followed, or sometimes accompanied, by a throbbing headache. Nausea is variable in degree and in the time it is experienced after the alcoholic beverage is taken. According to Morgan and Cagan, nausea will begin 30 to 60 min after the patient has taken 40 to 50 g of alcohol.[224] With the development of nausea, the patient becomes pale and may feel faint. A fall in blood pressure may occur but is by no means constant. Vague, unlocalized abdominal pain and vomiting may follow the development of nausea. Patients feel uneasy or actually frightened. Chest pain, which may be unilateral (usually left-sided) as well as vertigo and blurred vision may bring the patient to the doctor's office or to the accident room. Psychotic reaction may be induced by the drug, particularly in patients who have previously been schizophrenic.

When an alcoholic is taking Antabuse® he/she will develop this reaction, which is known as the acetaldehyde reaction, not only when alcoholic beverages are consumed, but also when food containing liquor, wine, or beer is eaten. Reactions may follow a very low level of alcohol intake. The severity of the reaction is related not only to the amount of alcohol taken, but also to the dose of the drug received.

Symptoms are due to inhibition of the enzyme aldehyde dehydrogenase such that there is a resultant increase in the blood levels of acetaldehyde. Acetaldehyde is a primary metabolite of alcohol.[225] Alcoholics will develop acetaldehyde reactions if they take medications containing alcohol including tinctures, tonics such as Geritol®, analgesic-sedatives such as Nyquil®, or certain cough syrups containing alcohol.

In nonalcoholics as well as in alcoholics, acetaldehyde-like reactions can be evoked by concurrent intake of metronidazole, a drug used in the treatment of vaginal Trichomonas infections, and alcoholic beverages or foods containing alcohol. Other drugs which produce an acetaldehyde-type reaction are citrated calcium carbimide used in Japan and Canada in the management of alcoholics, oral antidiabetic agents of the sulfonylurea group, and the antibiotic, chloramphenicol.[226]

Flushing reactions commonly occur when alcoholic beverages or foods are taken when someone is receiving tetrachloroethylene, which is an anthelmintic used particu-

TABLE 35

Tyramine Content of Foods and Beverages

| Foods/beverages | Tyramine content ($\mu$g/g) |
|---|---|
| Cheese | |
| Cheddar | 120—1500 |
| Camembert | 20—2000 |
| Emmenthaler | 225—1000 |
| Stilton | 466—2170 |
| Processed | 26—50 |
| Brie | 0—200 |
| Gruyere | 516 |
| Gouda | 20 |
| Brick, natural | 524 |
| Mozzarella | 410 |
| Roquefort | 27—520 |
| Parmesan | 4—290 |
| Romano | 238 |
| Provolone | 38 |
| Cottage | 5 |
| Fish | |
| Salted, dried | 0—470 |
| Pickled herring | 3000 |
| Meat | |
| Meat extracts | 95—304 |
| Beef liver (stored) | 274 |
| Chicken liver (stored) | 100 |
| Vegetables | |
| Avocado | 23 |
| Fruit | |
| Banana | 7 |
| Alcoholic Beverages | |
| Beer and ale | 1.8—11.2 |
| Wines | 0—25 |
| Chianti | 25.4 |
| Sherry | 3.6 |

From Lovenberg, W., Some vaso- and psychoactive substances in food, in *Natural Toxicants in Food*, National Academy of Science, Washington, D. C., 1973, 172, 174.

larly in the treatment of hookworm. This reaction may occur in those who are engaged in the manufacture of this drug, since absorption of the drug may be through the skin.[227]

## 4. Hypoglycemic Attacks

Diabetics on sulfonylurea drugs (oral antidiabetic agents), such as tolbutamide, who take alcohol or alcohol-containing foods at the same time may develop signs of hypoglycemia. Hypoglycemia is associated with faintness, crying, irrational behavior, somnolence, and coma.[228]

## 5. Hypokalemia

Hypokalemia, or potassium deficiency, may result from excessive consumption of licorice. The active substance in the licorice is glycyrrhizic acid. Interactive drugs, potentiating the effect of licorice in causing potassium depletion, include alcohol and oral diuretics, particularly thiazide diuretics. Hypokalemia causes profound weakness and cardiac arythmias.[229]

It has also been reported that paraaminosalicylic acid (PAS) used in the treatment of tuberculosis causes hypokalemia and that the hypokalemic effect of this drug is potentiated by licorice eating.[230]

## B. Effects of Food on the Bioavailability of Drugs

Pharmaceutical tradition has dictated whether a drug should be given before or after food. Drugs given to fasting persons or before meals have been those used supposedly to stimulate appetite (tonics) and other drugs for which experience showed that the desired therapeutic effect was enhanced by having the drug taken when the patient was fasting rather than after food. In the past, drugs taken after food were usually those that were believed to cause gastric irritation if taken without food.

It is now known that the absorption of drugs may be increased or decreased by feeding, by foods, or by specific nutrients or nonnutrient components of the diet. Delayed absorption of the following drugs occurs when they are taken at the same time as food, though the total amount of drug absorbed appears not to be significantly altered. These drugs are amoricillin, cephalexin, sulfanilimide, sulfadiazine, sulfamethoxine, sulfioxazole, aspirin, acetaminophen, digoxin, and furosemide.[231,232] Delayed absorption of acetaminophen when food is taken appears to be due to the pectin content of the diet. Certain drugs are more efficiently absorbed when taken in the fasting state.[233] Impaired absorption of penicillin G and V as well, to a lesser extent, as ampicillin, amoxicillin, tetracycline, demethylchlortetacycline, methacycline, oxytetracycline, aspirin, propantheline (Probanthine®), levodopa (L-dopa), rifampicin, doxycycline, isoniazid, and phenobarbital occurs when these drugs are actually taken at meal times.[234] Absorption of tetracycline and demethylchlortetracycline is particularly reduced by concomitant drinking of milk or eating of milk products such as cheese. Calcium chelates of tetracycline are produced, which are not absorbed from the GI tract.[234] Pharmacological doses of iron salts, such as ferrous sulfate, also reduce tetracycline absorption.[235]

Absorption of a number of drugs, on the other hand, may be promoted by food intake. The absorption of griseofulvin, nitrofurantoin, proproxyphene, lithium salts, hydralazine, propranolol, metoprolol, spironolactone or its primary metabolite, canrenone, carbamazepine, and hetacillin, is enhanced by food intake.[236-240] The possible reasons for this effect are the increased degree of dissolution of drug tablets by prolonged contact with gastric contents, and improved absorption of drugs brought about by delayed gastric emptying. Food factors which influence drug absorption need further definitions; however, Table 36 summarizes present knowledge of drugs which are absorbed more rapidly or more efficiently in the fasting state as well as drugs which are better absorbed when taken with food.

## C. Drug-Induced Malnutrition

Mechanisms whereby drugs can affect nutritional status are listed below.

## 1. Effects on Food Intake

Drugs can alter food intake by causing a perversion or loss of appetite, because interest in food is lost due to sedation, or because the drugs cause gastrointestinal

TABLE 36

Food Effects on Drug Absorption

Delayed Reduced

**Drugs for Which Absorption is Delayed or Reduced by Food**

| | |
|---|---|
| Amoxicillin | Penicillin G |
| Cephalexin | Penicillin V (K) |
| Cephradine | Phenethicillin |
| Sulfanilamide | Ampicillin |
| Sulfadiazine | Amoxicillin |
| Sulfadimethoxine | Tetracycline |
| Sulfamethoxypyridazine | Demethylchlortetracycline |
| Sulfisoxazole | Methacycline |
| Sulfasymazine | Oxytetracycline |
| Aspirin | Aspirin |
| Acetaminophen[a] | Propantheline |
| Digoxin | Levodopa |
| Furosemide | Rifampicin |
| Potassium ion | Doxycycline |
| | Isoniazid |
| | Phenobarbital |

**Drugs Which are Better Absorbed with Food**

| | |
|---|---|
| Griseofulvin[b] | Hydralazine[c] |
| Nitrofurantoin — macrocrystalline and microcrystalline | Propranolol[c] |
| Propoxyphene | Metoprolol[c] |
| Riboflavin | Spironolactone[c] |
| Riboflavin-5'-phosphate | Carbamazepine |
| Lithium citrate | Hetacillin |

[a]  Probable pectin effect.
[b]  Fat promotes absorption.
[c]  Food promoting bioavailability included Swedish crisp bread high in dietary fiber.

Modified from Welling.[231]

disturbances whenever food is taken. Those affecting appetite may have a central or a peripheral effect. Appetite-reducing drugs which function centrally are the unsubstituted or substituted amphetamines. Whereas these drugs do not have a profound or lasting effect in diminishing food intake in obese people, where intake of food is highly influenced by its sensory appeal, amphetamines do reduce appetite and hence, food intake when taken either as an illicit drug, or when given to children.

Dextroamphetamine, which has been used as a maintenance drug in the management of hyperactivity in children, has been shown to impair appetite and to reduce the rate of weight gain and growth in these youngsters. Return to normal appetite occurs if the amphetamine administration is discontinued in these children, and catch-up growth follows. Amphetamine abusers' emaciation may result from low energy consumption. Over-the-counter weight-reducing drugs, which may contain bulking agents such as methylcellulose, or mild stimulants such as phenylpropanolamine, an ephedrine-like drug, do not have a substantial effect on food intake and cannot be considered as the drugs of choice in the management of obesity.

Any drug which induces nausea is likely to reduce food intake and hence, contribute

TABLE 37

Drugs and Drug Groups Which Affect Food Intake

### Drugs Which Reduce Food Intake

| | |
|---|---|
| Intentional appetite-reducing drugs | Amphetamines |
| | Methylphenidate |
| | Mazindol |
| | Bulk agents |
| Drugs which produce aversive responses to food | Digitalis and related alkaloids |
| | Cancer chemotherapeutic agents |
| | Anticonvulsants |
| | Alcohol (excess) |
| | Sedatives and tranquilizers (excess) |

### Drugs Which Increase Food Intake

| | |
|---|---|
| Intentional appetite-stimulating drugs | Cyproheptadine |
| | Alcohol (small amount) |
| Drugs with side effects of increasing appetite | Corticosteroids |
| | Oral hypoglycemic agents |
| | Insulin |
| | Psychotropic drugs — phenothiazines and benzodiazepines |

to weight loss. Digitalis, which is usually given for long periods of time, if also prescribed and taken at a high-dosage level, can cause severe wasting or cachexia, because nausea diminishes the appeal of food.

Cancer chemotherapeutic drugs usually cause loss of appetite. Reduction in food intake associated with periods of active anticancer or antileukemia drug use is due (1) to alteration in the taste of food, (2) to sore mouth or throat, (3) the drug may cause postprandial abdominal pain or diarrhea, and (4) the systemic effects of the drug may be such as to produce severe malaise.

Phenothiazine and benzodiazepine tranquilizers have a variable effect on appetite, and food intakes are dependent on the circumstances in which they are taken. When phenothiazine or benzodiazepine tranquilizers are given to psychotic patients, particularly those with agitated depression, food intake is improved because the level of agitation is reduced or controlled. Large appetites in institutionalized psychotic patients on these drugs may be a cause of obesity both because food intake is increased and also because physical activity may be curtailed. On the other hand, when tranquilizers are given in large doses to elderly patients such that their level of consciousness is reduced, then food intake may be inadequate.

Drugs other than phenothiazine or benzodiazepine tranquilizers, which can enhance food intake through change in appetite, include cyproheptadine (Periactin®). Cyproheptadine has been employed to encourage increases in food intake among debilitated and malnourished groups or patients.[11] Major drugs or drug groups having an effect on food intake are tabulated in Table 37.

### 2. Drug-Induced Malabsorption

Drugs can cause malabsorption and therefore reduce nutrient gain when food is taken. Mechanisms whereby malabsorption is brought about by drugs include (1) reduced secretion or release of pancreatic digestive enzymes, leading to maldigestion and hence malabsorption, (2) damage to the brush border and/or the mucosa of the small intestine such that nutrient transport across the intestinal membrane is reduced, (3)

TABLE 38

**Drugs Which Cause Malabsorption**

| Drug group | Drug | Proposed mode of action | Nutrient losses |
|---|---|---|---|
| Laxatives | Mineral oil | Solubilization of nutrients; Micelle formation | Vitamin A,D,K |
| | Phenolphthalein, bisacodyl | Loss of intestinal mucosal integrity? | Potassium ($K^+$), calcium ($Ca^{++}$) |
| Antibacterial agents | Neomycin | Precipitation of bile salts; inhibition of pancreatic lipase; villous damage | Fat, nitrogen (N), $Ca^{++}$, Fe, $Na^+$, $K^+$, disaccharides, vitamin $B_{12}$ |
| | Paraaminosalicylic acid | Ileal transport inhibition | Fat, vitamin $B_{12}$ |
| Antiinflammatory drugs | Colchicine | Villous damage | Fat, N, $Na^+$, $K^+$, vitamin $B_{12}$ |
| | Sulfasalazine | — | Folacin |
| | Mefenamic acid | Unknown | Fat |
| Antacids | Aluminum hydroxide, magnesium hydroxide | Formation of aluminum and magnesium phosphate | Phosphorus |
| Hypolipedemic agents | Cholestyramine | Adsorbs nutrients; removes bile acids required in vitamin absorption | Folacin, vitamins A,D,E,K, |
| Anticonvulsants | Diphenylhydantoin, phenobarbitone, glutethimide | Impaired production of vitamin D-dependent calcium-binding protein; decreased intestinal transit of folacin | $Ca^{++}$, folacin |

drugs may have a selective effect on active transport mechanisms, (4) nutrients may be solubilized in a drug vehicle, (5) nutrients may be converted to an insoluble form by drugs so that absorption can no longer take place, and (6) drugs may adsorb nutrients.

Drugs which are in common use that cause nutrient malabsorption include antacids, laxatives including mineral oil and phenolphthalein, the antibiotic neomycin, and colchicine used in the treatment of gout. Cholestyramine, now employed extensively as a drug to control blood cholesterol levels in patients with hyperlipoproteinemias of genetic origin, causes malabsorption of fats and fat-soluble vitamins because the drug adsorbs bile acids which are required for the optimal absorption both of fats and of fat-soluble vitamins. In addition, cholestyramine causes malabsorption of folacin, because this B vitamin is adsorbed to the resin in the lumen of the intestine.[241] Drugs which cause malabsorption and their mode of action are summarized in Table 38.

### 3. Drugs Causing Mineral Depletion
#### a. Potassium

Two major groups of drugs can lead to low potassium levels in the serum (hypokalemia) with clinical signs of potassium deficiency. Abuse of laxatives or cathartics causes excessive potassium loss via the intestine. Patients who are laxative abusers may be elderly or middle-aged persons with obsessions about bowel regularity or young women with anorexia nervosa.[242] A common laxative drug producing hypokalemia is phenolphthalein, present in several over-the-counter preparations and the active ingredient of Ex-Lax®.

Oral diuretics including thiazide drugs, chlorthalidone (Hygroton®), ethacrynic acid (Edecrin®), and furosemide (Lasix®) can all produce potassium depletion.[243] Risk of severe potassium depletion with intake of diuretics is greatest for patients on low potassium intake or those whose potassium excretion is increased due to alcohol abuse or alcoholic cirrhosis.[244]

A third group of drugs which may produce potassium depletion are the corticosteroids. Risk is related to the mineralocorticoid properties of particular drugs within this group. However, muscle weakness, which is a well-recognized side effect of high dosage and prolonged corticosteroid therapy is more likely to be due to muscle wasting than to potassium depletion.[245] Potassium depletion induced by diuretics is a major cause of digitalis intoxication.

### b. Magnesium and Zinc

The commonest drug affecting magnesium and zinc excretion is alcohol. Increased excretion of magnesium and zinc occurs in acutely drinking alcoholics, and in those patients the urinary losses of these elements may contribute to the development of magnesium and zinc deficiency. Contributory causes of magnesium deficiency in alcoholics are low intake and use of diuretics. Zinc and magnesium deficiency in alcoholics is also related to low-intake, long-term intravenous feeding in alcoholic brain syndromes without addition of zinc to the infusion fluid, muscle wasting after injury or surgery with loss of tissue zinc, as well as use of diuretics.[204,246]

Oral diuretics increase urinary losses of magnesium and zinc, and depletion may result either from use of these diuretics alone or combined intake of digitalis glycosides and oral diuretics.[247,248]

### c. Calcium

Losses of calcium in the urine are increased by oral diuretics including furosemide, ethacrynic acid, and triamterene.[249] Calcium malabsorption is also associated with intake of corticosteroids. Calcium malabsorption is commonly secondary to drug-induced vitamin D deficiency which may be induced by diphenylhydantoin, phenobarbital, and glutethimide.[250,251]

### d. Phosphate

Antacids containing magnesium and aluminum hydroxides can cause phosphate malabsorption and depletion because dietary phosphates are excreted in the feces as magnesium and aluminum salts. The risk of severe phosphate depletion occurs with antacid abuse which is common in alcoholics. Symptoms include malaise and muscle weakness. With severe antacid-induced phosphate depletion, osteomalacia may supervene.[252,253]

Drugs which cause mineral wasting and deficiency either by urinary or fecal loss are indicated in Table 39.

### 4. Vitamin Antagonists

Drugs may function as antinutrients or vitamin antagonists, and in certain disease situations drugs are deliberately used for this purpose. For example, the antifolate (anti-folacin) effects of methotrexate are utilized in the treatment of leukemia and other forms of malignant disease. Folate antagonists ar drugs which bind to the dihydrofolate reductase enzyme which is required for the conversion of the vitamin to its active coenzyme form. Drugs other than methotrexate which have an antifolate effect

TABLE 39

**Drugs Which Cause Mineral Wasting and Deficiency**

| Drug group | Drug | Mineral wasted | Route of mineral loss | |
|---|---|---|---|---|
| | | | Urine | Feces |
| Laxatives | Phenolphthalein Bisacodyl | Potassium | − | + |
| Diuretics | Furosemide | Potassium, calcium | + | − |
| | Ethacrynic acid | Potassium, magnesium, zinc | | |
| | Thiazides | Potassium, magnesium, zinc | | |
| | Triamterene | Calcium, magnesium, zinc | + | |
| Cardiac glycosides | Digitalis and related drugs Strophathin | Calcium, magnesium | + | |
| Hormones | Cortisone and related drugs | Calcium Potassium | + | |
| Chelating agents | D-penicillamine EDTA | Zinc, copper Zinc | + | |
| Antacids | Magnesium and aluminum hydroxide | Phosphate | | + |
| Anticonvulsants | Diphenylhydantoin Phenobarbital | Calcium | | ± |
| Alcoholic beverages | Ethanol | Magnesium Zinc | + | |

include pyrimethamine, used in the prevention and control of malaria, and triamterene, which is a diuretic. Anticoagulants, which are used to treat or prevent thrombophlebitis, are antivitamin K compounds. Massive intake of vitamin K inhibits their effectiveness.

Certain drugs have antinutrient properties which present as unwanted side effects. For example, isoniazid, which has been widely used in the treatment of tuberculosis, has an antivitamin $B_6$ effect. The major resultant effect is the development of neuritis.[90] Antinutrient effects of drugs and whether or not these effects are intentional or incidental are indicated in Table 40.

## 5. Drug-Induced Nutritional Deficiencies of Uncertain Causation

Anticonvulsant drugs including diphenylhydantoin, phenobarbital, and primidone, cause deficiencies of vitamins D, K, and folic acid. These deficiencies occur with prolonged usage of the drugs by those whose status is marginal with respect to these nutrients. The pathophysiological effects of anticonvulsant drugs on nutrient absorption and metabolism are not presently well understood, though evidence points to an increased rate of metabolism of vitamins D, K, and folic acid. Further, studies have shown that diphenylhydantoin can decrease folacin absorption as well as calcium absorption. Prenatal or perinatal malnutrition can also be induced by these drugs. For example, neonatal vitamin K deficiency may occur in the infants of epileptic women who have received anticonvulsant drugs such as diphenylhydantoin during pregnancy. Malformations associated with the fetal diphenylhydantoin syndrome are probably due to direct toxic effects of the drug on the fetus during embryogenesis, though fetal

TABLE 40

Drugs Which Function as Antinutrients

**Folate antagonists**
Methotrexate (intentional)
Pyrimethamine
Triamterene
Pentamidine isethionate
Trimethoprim

**Vitamin B₆ antagonists**
Isonicotinic acid hydrazide
Hydralazine
Cycloserine
Pyrazinamide
Ethionamide
Penicillamine
L-dopa

**Vitamin K antagonists**
Warfarin
Acenocoumarol
Phenprocoumon

From Roe, D. A., Effects of drugs on nutrition, *Life Sci.*,
15, 1219, 1974.

TABLE 41

Adverse Effects of Anticonvulsant
Drugs

Anorexia
Folacin deficiency
Rickets and osteomalacia
Coagulation disorders
 (Vitamin K deficiency in
 newborn infants)
Fetal malnutrition

malnutrition has not been excluded. Folacin deficiency in epileptics on anticonvulsant drugs is related in part to impairment in food intake which may also mean that food folacin intake is reduced.

High doses of folic acid when given to epileptics on diphenylhydantoin and pheno-barbital can decrease the effectiveness of these therapeutic agents by lowering blood levels of the drug. For this reason, prevention of folacin deficiency in epileptics should be by administration of vitamin supplements in physiological dosages.[254] Adverse nutritional effects of anticonvulsant drugs are shown in Table 41.

Nutritional requirements are altered in women on oral contraceptives. Depletion of B vitamins including folacin, riboflavin, and vitamin B₆ as well as vitamin C have been identified by biochemical measurements in women on oral contraceptives (Table 42). Clinical evidence of nutritional deficiency is rare and has only been found in a few women who have also been on deficient diets or who have malabsorption syndromes independently of oral contraceptive use. Mechanisms whereby contraceptive steroids induced B vitamin deficiency or depletion, or depletion of vitamin C, are not under-

TABLE 42

Nutritional Aberrations Attributed to Contraceptive Steroids

| Nutrient | Effect |
| --- | --- |
| Folacin | Serum level decreased |
| | Erythrocyte level decreased |
| | Megaloblastic anemia (rare) |
| Vitamin $B_{12}$ | Serum level decreased |
| Riboflavin | Erythrocyte level decreased |
| | Glossitis (rare) |
| Vitamin $B_6$ | Disturbed tryptophan metabolism |
| | Plasma PLP decreased |
| | Depression |
| Ascorbic acid | Leukocyte content decreased |
| | Platelet level decreased |
| Vitamin A | Plasma level decreased |
| Iron | Serum level increased |
| | TIBC increased |
| Copper | Plasma copper increased |
| | Ceruloplasmin increased |
| Zinc | Plasma zinc decreased |

From Roe, D. A., Nutrition and the contraceptive pill, in *Nutritional Disorders of American Women*, Winick, M. D., Ed., John Wiley & Sons, New York, 1977.

stood. Earlier human studies suggesting that contraceptive steroids induce folacin malabsorption have not been substantiated. More recent studies suggest that the formulation of the oral contraceptive is important and that folacin depletion may occur particularly with use of preparations containing mestranol.

Blood levels of other nutrients may be altered by the Pill, and it has been shown that these drugs cause a significant increase in blood levels of vitamin A.

Nutritional problems which have been attributed to contraceptive steroids are related to undernutrition or overnutrition of the community or population from which the sample is drawn. In subgroups of the population in which obesity is prevalent, obesity will be commoner in women on oral contraceptives. However, since high estrogen contraceptive steroids have been discontinued, causal relationships between weight gain and intake of these drugs may be questioned. Women who are underweight are more subject to side effects of some oral contraceptives than those who are of normal weight or above normal weight. Iron deficiency anemia is less likely to occur in women on oral contraceptives than those who are not, because of reduced menstrual blood flow when these drugs are taken.[255]

## D. Diet, Nutrition, and Drug Metabolism

In laboratory animals as well as in human subjects, it has been shown that changes in the diet as well as changes in nutritional status can affect the rate of drug metabolism. Model drugs including antipyrine and theophylline are metabolized faster when a person is on a high protein-low carbohydrate diet rather than a high carbohydrate-low protein diet. These effects of diet show intersubject variability.[256] It is presently assumed that effects of protein level in the diet on the rate of drug metabolism apply to other drugs. Nonnutrient dietary components including indolic compounds present in cabbage and brussels sprouts may speed up drug metabolism. Cooking methods also alter the rate of drug metabolism, particularly charcoal broiling which can promote conversion of drugs to the proximal metabolite. Changes in the diet, as for example change from a meat to a nonmeat diet or from a low- to a high-fiber diet will alter the range of microflora in the gastrointestinal tract and influence the rate of drug metabolism or the range of drug metabolites produced.

Toxicity of drugs is influenced by nutritional status. For drugs normally bound to serum albumin, hypoalbuminemia results in more free drug in the body, which may explain an increased frequency of side effects. Side effects of prednisone are increased in patients with protein deficiency for this reason.[257]

Protein depletion and protein-energy malnutrition affect drug toxicity. Conversion of drugs to their metabolites by hydroxylation and/or conjugation is slowed in states of protein malnutrition. However, whether this leads to enhanced or decreased drug toxicity depends on whether the primary drug or its metabolite is the more toxic compound. Studies of chloramphenicol metabolism in children with protein-energy malnutrition have shown that there is a slower rate of drug biotransformation in the malnourished children, and that these children clear the drug more slowly from the plasma.[258] Drug dosage in malnourished persons needs to be adjusted both to body weight and with respect to changes in the rate of drug metabolism. Vitamin C deficiency may also inhibit drug metabolism.[259]

## SECTION VI
## NUTRITIONAL CARE IN THE HEALTH DELIVERY SYSTEM: A TEAM APPROACH

The goals of nutritional care are preventive and therapeutic. Aims of preventive nutritional care can be summarized under the following headings:

### A. Community Nutrition — Goals, Programs, and Personnel

1. Nutrition education
   a. Create public awareness as to nutritional needs for health
   b. Disseminate information on relationships between prevailing food habits and the development of chronic or degenerative diseases
   c. Explain nutritional risks involved in food, drug, and alcohol abuse
   d. Educate health professionals to risks of malnutrition in special population groups
   e. Upgrade nutrition attitudes of health professionals
   f. Initiate and evaluate programs to improve nutrition of specific target groups varying in age, education, and physiological status
   g. Develop nutrition counseling services
2. Nutrition policy
   a. Establish liaison with the food industry, so that food product development would be based on nutritional needs
   b. Demand that public service and commercial television give time to nutrition programs
   c. Adopt national dietary goals (Table 43)
   d. Monitor safety and efficacy of all foods and food supplements
   e. Create socioeconomic conditions which make a nutritionally adequate and health-promoting diet available to all
3. Nutrition research
   a. Determine how best nutrition information can be communicated to the public in order that they will modify their dietary habits
   b. Assess optimum techniques for assessment of people at nutritional risk in community and clinic situations
   c. Develop specific roles for health professionals in delivery of nutrition services
   d. Evaluate the most economic allocation of health care personnel to community nutrition services
   e. Examine effects of introducing nutrition into the curriculum of health professional training schools with respect to change in nutrition attitudes, eating practices, and nutritional status of graduates
   f. Long-range assessment of nutrition training of health professionals on community nutrition practices and status
   g. Determine the value of short nutrition courses and/or workshops in continuing education programs for health professionals

Current barriers to effective nutrition education and/or other preventive nutrition services are competition from the media, competition from the food industry, indecision on nutrition goals, and erroneous belief by health professionals that delivery of nutrition information, regardless of teaching method or quality, will result in change in food habits of the target group.

Hollingsworth remarked " . . . the health of many people could be improved if they

TABLE 43

**U.S. Dietary Goals**

1.  Increase carbohydrate consumption to account for 55 to 60% of the (caloric) intake.
2.  Reduce overall fat consumption from approximately 40 to 30% of energy intake.
3.  Reduce saturated fat consumption to account for about 10% of total energy intake; and balance that with polyunsaturated and monounsaturated fats, which should account for about 10% of energy intake each.
4.  Reduce cholesterol consumption to about 300 mg a day.
5.  Reduce sugar consumption by about 40% to account for about 15% of total energy intake.
6.  Reduce salt consumption by about 50 to 85% to approximately 3 g a day.

**The Goals Suggest the Following Changes in Food Selection and Preparation**
1.  Increase consumption of fruits and vegetables and whole grains.
2.  Decrease consumption of meat and increase consumption of poultry and fish.
3.  Decrease consumption of foods high in fat and partially substitute polyunsaturated fat for saturated fat.
4.  Substitute nonfat milk for whole milk.
5.  Decrease consumption of butterfat, eggs, and other high cholesterol sources.
6.  Decrease consumption of sugar and foods high in sugar content.
7.  Decrease consumption of salt and foods high in salt content.

From U. S. Dietary Goals, Government Printing Office, Washington, D. C., 1977, and from Latham, M. C. and Stephenson, L. S., *J. Nutr. Educ.*, 9, 152, 1977.

TABLE 44

**Nutrition Responsibilities of Non-nutritional Health Care Professionals**

Nutrition education
Nutritional Assessment
Preventive nutritional care
Nutrition intervention
  Community
  Hospital

were to reduce their consumption of fats (particularly saturated fats), sugar and alcohol and increased their consumption of vegetables, fruit, bread and other cereal foods, particularly whole grain ones. Changes on these lines would run counter to many recent trends in food consumption, and also to much former nutrition education. How are we to achieve a concensus of opinion and subsequent action?"[260]

We believe that preventive nutritional care must be offered by health professionals having the most daily contact with the public. We identify responsibility for preventive nutrition care as belonging to health educators in schools, physicians, nurse clinicians, nurses and physicians' assistants in family practice clinics, dentists and dental hygienists, staffs of family planning clinics, and most importantly, pharmacists. Guidelines on nutrition responsibilities of these health professionals are listed in Table 44. These guidelines for nutrition services to meet the need are based on recommendations set forth in the Hearings Before the Select Committee on Nutrition and Human Needs of the U.S. Senate, October 1977. In a detailed statement before this Committee, Dr. Julius B. Richmond, Assistant Secretary for Health, Department of Health, Education, and Welfare, stated:

Nutrition has long been an important component of the public health service and the Department of

Health, Education, and Welfare . . . Society would reap considerable benefit from dietary and other life-style modification, especially if these are combined with more effective use of our health care resources and environmental changes aimed at promoting health. People want and could benefit from more definitive and practical guidance about nutrition and its relation to health and disease. Consumers already have responded to information and education about nutrition and diet . . . More of the existing knowledge about nutrition and health must be effectively communicated to the public, and particularly to target populations often neglected by current activities.[261]

A team approach to preventive nutritional care implies definition and specific roles for health professionals and also that health professionals have sufficient commitment to and knowledge of nutrition. Health professionals, whose personal attitudes to nutrition and diet are unsatisfactory, are not going to be able to motivate their patients or public to improve their eating practices. Priorities should be set for training of health professionals in nutrition who are good role models. Further, there is a specific need for health workers to understand and to be able to communicate health risks of dietary indiscretion. The requirement that health workers should have a basic knowledge of nutrition implies that texts are available on nutrition which are understandable by those who are not nutrition specialists. There is also an urgent need to include nutrition courses into the training programs of health workers.

The American Dietetic Association has made the following recommendations for the nutrition component of health services delivery systems:[262,263]

1. That nutrition services be under the supervision of qualified nutrition personnel, and that such services should be a component of all health and health-related programs
2. That nutrition services should be designed to reach the total population, but with priority being given to nutritionally vulnerable groups
3. That nutritional care as a part of health care should be available to everybody on a continuing and coordinated basis
4. That nutritional care should be integrated into preventive as well as diagnostic, curative, and restorative health services
5. That the planning and administration of nutritional care should be under the direction of those who are professionally educated in nutrition, particularly as nutrition relates to human health needs
6. That comprehensive health care plans include a nutritional care component with realistic and appropriate staffing patterns
7. That services supplied by dietitians should be expanded by inclusion of supportive personnel
8. That adequate funding for preventive as well as curative nutritional care services should be provided at the federal or local level
9. That the economic benefits of nutritional care in a comprehensive health care system should be examined
10. That any national health insurance program should include incentives for the development of preventive services including nutritional services
11. That standards for nutritional care services be included in federal and state guidelines and regulations for health care delivery[264]

Clinic systems which provide health care services to nutritionally vulnerable groups offer a captive and target audience for nutrition education and counseling. The clinic situation is one which lends itself to nutritional care by the team approach. The team may include physicians, nurse clinicians, nurses, a clinical or community nutritionist, dentists, pharmacists, social workers, physicians' assistants, and other health paraprofessionals. Whenever a clinical nutritionist is a member of the clinic staff, he/she will share the responsibility for nutritionists' assessment of patients and determination of nutritional needs of patients with physicians and/or dentists. When clinics do not have a nutritionist on the staff, a nurse clinician with nutrition training may share with the physician duties pertaining to nutritional assessment and determination of nutritional needs of the clinic population.

Alternate strategies for nutrition education and counseling of patients in clinics are by direct services, by a clinical or community nutritionist, by training of clinic staff to perform these functions, and by provision of nutrition education materials. The clinic staff may be trained at the clinic site to recognize nutritional and food-related problems of the patients and to offer specific types of nutritional advice, as well as in making referrals of patients to nutrition support services in the community. Success of the nutrition program in a clinic demands that a coordinator of nutritional services be appointed, who is usually the nurse manager. This individual should be given responsibility for arranging the nutrition education of clinic nurses and health professionals, for purchase of nutrition education materials, and for maintaining the quality of nutritional counseling and follow-up. If the clinic has a social worker, he/she would be the key person to make referral of patients to nutrition programs in the community and should define eligibility of patients for special food programs. The latter activity will require coordination between the local Department of Social Services and the clinic. The most important aspect of any nutrition service in a clinic is that compliance of patients with a prescribed nutrition program or therapy must be examined and monitored.

### 1. Specific Nutrition Responsibilities of Health Professionals
#### a. Physicians

**Family practice physicians** — Family practice physicians working in clinics, HMO's, or in private offices should (1) be able to answer patients' questions about their nutritional needs and how these can be supplied by diet, (2) be able to carry out simple nutritional assessment including dietary evaluation, (3) be able to make a nutritional needs assessment of patients of different ages, (4) identify patients at high nutritional risk, (5) know when patients need extensive nutritional evaluation and work-up, (6) be able to give dietary guidelines for patients with acute health problems and with common chronic diseases including atherosclerotic heart disease, hypertension, diabetes, peptic ulcer, constipation, diverticulosis, and congestive heart failure, (7) should be able to advise patients on avoidance of unsound and dangerous fad diets, (8) should recognize and prevent diet-drug interactions and incompatibilities, (9) should recognize anorexia nervosa, (10) should be able to explain the prudent diet and other dietary modifications which can prevent disease, (11) should be able to recognize malnutrition as a result of child abuse, alcoholism, drug addiction, and geriatric disability, and (12) should be able and willing to make referrals of patients to clinical or community nutritionists and/or to nutritional programs in the community.

**Obstetrician** — Obstetricians should (1) advise patients of the nutritional needs of normal pregnancy and lactation, (2) guide patients on appropriate weight gain or loss during these periods of physiological stress, (3) diagnose and treat nutritional anemias of pregnancy, (4) understand nutritional risk of the fetus to maternal alcoholism, drug addiction, smoking, and intake of over-the-counter and prescription medications, and (5) recognize patients including pregnant adolescents at high nutritional risk and make referrals of such patients for nutritional counseling.

**Pediatricians** — Pediatricians in clinic or office practice should (1) promote breast-feeding, (2) advise patients on the feeding of infants and children, (3) recognize and counsel families on dietary mismanagement, (4) recognize and treat common food allergies, (5) recognize and treat undernutrition or malnutrition and understand underlying causes. Wherever malnutrition is identified in the child and the cause is not readily identified, patients should be referred for diagnostic work-up at a hospital or medical center, and (6) counsel parents on prevention of childhood obesity, determine the cause, and make recommendations for nutritional management.

### b. Dentists

Dentists as well as dental hygienists should (1) inform patients and their families about relationships between diet and tooth decay, (2) convey information on nutritional needs of pregnant women and the neonate with respect to tooth development, and (3) be able to influence acceptance of the fluoridation of water supplies and/or fluoride treatments to prevent tooth decay.

Orthodontists should educate patients about relationships of diet, particularly sugar intake, to plaque formation.

Specialists in restorative dentistry should be able to analyze patients' diets, should explain the dietary causes of tooth decay, and should counsel patients to eliminate foods which are cariogenic.

Oral surgeons must prepare patients for oral surgery by nutritional support, and must promote healing in the postoperative period by prescription of appropriate diet and nutritional supplements.

Dentists and dental specialists should (1) advise new denture wearers about a nutritious soft diet during the period of adjustment, (2) advise such patients on the need for rapid resumption of a balanced, solid diet, and (3) must educate elderly patients with masticatory problems on the health risks of avoiding solid foods of high nutritional value.

### c. Public Health Nurses

Public health nurses should (1) be able to detect malnutrition in elderly and housebound or bedridden patients, (2) be able to assess compliance of patients on special diets, (3) be responsible for outreach prenatal and postnatal clinics, (4) recognize nutritional neglect of infants and children, and (5) make referral of patients with obvious or suspected nutritional problems through community nutrition programs, and/or clinics, and/or hospitals. Other nutrition-related responsibilities of nurses are discussed under the section on nutrition support services in hospitals.

### d. Pharmacists

Clinical and community pharmacists should serve as resource persons for nutrition information, particularly since they may be the first health professional approached by a patient in need of health advice. They should (1) be able to explain and counsel patients on food-drug interactions and incompatibilities, as well as drug-induced nutritional deficiencies, (2) knowing vitamin-mineral requirements (RDA), they should be able to advise patients on appropriate intake of OTC vitamin-mineral preparations, (3) they should be able to tell patients on request the composition of vitamin preparations or other nutritional supplements which may contain allergens, responsible for common hypersensitivities, (4) they must educate patients to the dietary causes of constipation and advise against the use of laxatives except as prescribed under careful supervision, (5) they must be constantly alert to the risks of new and established fad diets and diet foods and should advise patients of these risks, (6) they should be able to communicate with physicians on composition of nutrient supplements and formula feeds and composition of infant formulas including soybean formulas and nutrient mixtures intended for infants and children with metabolic disorders, (7) they should know the alcohol content of medications and discourage use of such medications by alcoholics, (8) they should label prescription drugs to make it clear to patients when the drug should be taken in relation to food, (9) they should promote diet as well as drug adherence in patients with heart disease, diabetes, hypertension, and peptic ulcer, and (10) they should be able to educate patients and to remind health professionals of the risks of vitamin intoxication, acute iron overload, and health foods which may contain heavy metal or other contaminants.

Nutrition education for the health professional and the interactive roles of health professionals in supplying nutrition information to one another and to their patients is described in the Proceedings of the 1977 Conference on Nutrition Education for the Health Professionals, edited by T. E. Burford and B. E. Sanders (SUNY Buffalo, 1978). Copies of this report can be obtained from the Educational Communications Center, State University of New York at Buffalo, Room 1, Foster Annex, Main Street Campus, Buffalo, NY 14214.

## B. Nutritional Support in Hospitals

A nutrition support team should be available in community hospitals as well as in medical centers. The team should include one or more physicians experienced in the nutritional management of acute and life-threatening disease and the appropriate indication for defined formula diets, tube feeding, and total parenteral alimentation. Other key members of the team should be the hospital dietitian and/or clinical nutritionist, a clinical pharmacist, a senior member of the nursing staff, and a hospital social worker. Leadership of the team may rotate, but in the absence of a nutritionist, it should be appropriately accepted by the pharmacist or the nurse who would then play a coordinating role. Physicians and surgeons in the hospital should be advised to obtain team advice and assistance in nutritional assessment of patients, nutritional management of patients, nutritional support of special patient groups including those with injuries, burns, diseases requiring major surgical intervention, as well as those with chronic liver and kidney disease with functional impairment. The team should be responsible for hyperalimentation, for adult and pediatric cases, and a member of the team, usually the pharmacist, should be responsible for making up parenteral solutions under sterile conditions and for maintaining quality control of these solutions.

The nutrition team and/or one of its members should be available on request by physicians or surgeons to offer special advice on the nutritional needs and appropriate nutritional therapy of patients under their charge.

# REFERENCES

1. **Cosper, B. A. and Wakefield, L. M.,** Food choices of women, *J. Am. Diet. Assoc.,* 66, 152, 1975.
2. **McKenzie, J.,** The impact of economic and social status on food choice, *Proc. Nutr. Soc.,* 33, 67, 1974.
3. **Grotkowski, M. L. and Sims, L. S.,** Nutritional knowledge, attitudes and dietary practices of the elderly, *J. Am. Diet. Assoc.,* 72, 499, 1978.
4. **Cardenas, J., Gibbs, C. E., and Young, E. A.,** Nutritional beliefs and practices in primigravid Mexican-American women, *J. Am. Diet. Assoc.,* 69, 262, 1976.
5. **Greaves, J. P. and Berry, W. T. C.,** Medical, social and economic aspects of assessment of nutritional status, *Bibl. Nutr. Dieta,* 20, 1, 1974.
6. **Wilson, N. L. and Wilson, R. H. L.,** Dysnutrition and Boredom, in *Progress in Human Nutrition,* Margen, S., Ed., AVI Publishing, Westport, Conn., 1971, 111.
7. **Brown, P. T. and Bergan, J. G.,** The dietary status of practicing macrobiotics: a preliminary communication, *Ecol. Food Nutr.,* 4, 103, 1975.
8. **Patek, A. J., Toth, I. G., Saunders, M. G., Castro, A. M., and Engel, J. J.,** Alcohol and dietary factors in cirrhosis. An epidemiological study of 304 alcoholic patients, *Ann. Intern. Med.,* 135, 1053, 1975.
9. **Frankle, R. T.,** The nutritionist in a drug rehabilitation center: a medical school reaches out, *J. Am. Diet. Assoc.,* 65, 562, 1974.
10. **Mont, F. G.,** Undernutrition, in *Modern Nutrition in Health and Disease,* 2nd ed., Wohl, M. G. and Goodhart, R. S., Eds., Lea & Febiger, Philadelphia, 1968, 984.
11. **Pawan, G. L. S.,** Drugs and appetite, *Proc. Nutr. Soc.,* 33, 239, 1974.
12. **Evans, E.,** Quality of diet received by the patient, *Proc. Nutr. Soc.,* 37, 71, 1978.
13. **Roa, D. B.,** Malnutrition and disease in the aged, *J. Am. Geriatr. Soc.,* 21, 362, 1973.
14. **Comstock, G. W. and Stone, R. W.,** Changes in body weight and subcutaneous fatness related to smoking habits, *Arch. Environ. Health,* 24, 271, 1972.
15. **Comstock, G. W., Shah, F. K., Meyer, M. B., et al.,** Low birth weight and neonatal mortality rate related to maternal smoking and socioeconomic status, *Am. J. Obstet. Gynecol.,* 111, 53, 1971.
16. **McMurray, W. C.,** *Essentials of Human Metabolism,* Harper & Row, New York, 1977.
17. **Davidson, S., Passmore, R., Brock, J. F., and Truswell, A. S., Eds.,** *Human Nutrition Dietetics,* Longman, New York, 1975, 15.
18. **Harper, H. A., Rodwell, V. W., and Mayes, P. A.,** *Review of Physiological Chemistry,* 16th ed., Lange Medical Publishing, Los Altos, Calif., 1977.
19. **Aurand, L. W. and Woods, A. E.,** *Food Chemistry,* AVI Publishing, Westport, Conn., 1973.
20. **Paul, A. A. and Southgate, D. A. T.,** *McCance and Widdowson's The Composition of Foods,* Her Majesty's Stationery Office, Elsevier/North Holland Biomed. Press, Amsterdam, 1976.
21. **Meneely, G. R. and Battarbee, H. D.,** Sodium and potassium, *Nutr. Rev.,* 34, 225, 1976.
22. **Cotlove, E. and Hogben, C. A. M.,** Chloride, in *Mineral Metabolism,* Vol. 2 (Part B), Comar, C. L. and Bronner, F., Eds., Academic Press, New York, 1962.
23. **Linkswiler, H. M.,** Calcium, in *Present Knowledge in Nutrition,* 4th ed., Nutrition Foundation, New York, 1976, 232.
24. **Gallagher, J. C. and Lawrence Riggs, B.,** Nutrition and bone disorders, *N. Engl. J. Med.,* 298, 193, 1978.
25. **Wacker, W. E. C. and Parisi, A. F.,** Magnesium metabolism, *N. Engl. J. Med.,* 278, 712, 1968.
26. **Wolfe, S. M. and Victor, M.,** The relationship of hypomagnesemia to alcohol withdrawal seizures and delirium tremens, *Ann. N. Y. Acad. Sci.,* 162, 973, 1969.
27. **Dodgson, K. S. and Rose, F. A.,** Sulfoconjugation and sulfohydrolysis, in *Metabolic Conjugation and Metabolic Hydrolysis,* Vol. 1, Fishman, W. H., Ed., Academic Press, New York, 1970, 239.
28. **Layrisse, M., Martinez-Torres, C., and Roche, M.,** Effect of interaction of various foods on iron absorption, *Am. J. Clin. Nutr.,* 21, 1175, 1968.
29. **Moore, C. V.,** Iron, in *Modern Nutrition in Health and Disease,* 5th ed., Goodhart, R. S. and Shils, M. E., Eds., Lea & Febiger, Philadelphia, 1973.
30. **Disler, P. B., Lynch, S. R., Charlton, R. W., et al.,** The effect of tea on iron absorption, *Gut,* 16, 193, 1975.
31. **Björn-Rasmussen, E.,** Iron absorption from wheat bread: influence of various amounts of bran, *Nutr. Metab.,* 16, 101, 1974.
32. **Boehne, J. W.,** Iodine nutriture in the United States, *Proc. Conf. Food and Nutrition Board,* National Academy of Science/National Research Council, Washington, D.C., 1970.
33. **Stewart, J. C. and Vidor, G. I.,** Thyrotoxicosis induced by iodine contamination of food — a common unrecognized condition, *Br. Med. J.,* 1, 372, 1976.

34. **Halsted, J. A., Ronaghy, H. A., Abadi, P., et al.,** Zinc deficiency in man, *Am. J. Med.,* 53, 277, 1972.
35. **Pories, W. J., Henzel, J. H., Rob, C. C., et al.,** Acceleration of wound healing in man with zinc sulfate given by mouth, *Lancet,* 1, 121, 1967.
36. **Prasad, A. S.,** Metabolism of zinc and its deficiency in human subjects, in Prasad, A. S., Ed., *Zinc Metabolism,* Charles C Thomas, Springfield, Ill., 1966, 250.
37. **Goolamali, S. K. and Comaish, J. S.,** Zinc and the skin, *Int. J. Dermatol.,* 14, 182, 1975.
38. **Morrison, S. A., Russell, R. M., Carney, E. A., and Oaks, E. V.,** Zinc deficiency: a cause of abnormal dark adaptation in cirrhotics, *Am. J. Clin. Nutr.,* 31, 276, 1978.
39. **Viller, R. W., Bozian, R. C., Hess, E. V., et al.,** Manifestations of copper deficiency in a patient with systemic sclerosis on intravenous hyperalimentation, *N. Engl. J. Med.,* 291, 188, 1974.
40. **Dallman, P. R.,** The nutritional anemias, in *Hematology in Infancy and Childhood,* Nathan, D. G. and Oski, F. A., Eds., W. B. Saunders, Philadelphia, 1974.
41. **Ulmer, D. D.,** Trace elements, *N. Engl. J. Med.,* 297, 318, 1977.
42. **Mertz, W.,** Fortification of foods with vitamins and minerals, *Ann. N. Y. Acad. Sci.,* 300, 151, 1977.
43. Food Fortification, Editorial, *WHO Chron.,* 26, No. 7303, 1972.
44. **Anon.,** *Recommended Dietary Allowances,* 8th rev. ed., Food and Nutrition Board, NRC/NAS, Washington, D.C., 1974, 3.
45. **Macdonald, I.,** The effects of dietary fiber: are they all good?, in *Fiber in Human Nutrition,* Spiller, G. A. and Amen, R. J., Eds., Plenum Press, New York, 1976, 263.
46. **Southgate, D. A. T.,** The chemistry of dietary fiber, in *Fiber in Human Nutrition,* Spiller, G. A. and Amen, R. J., Eds., Plenum Press, New York, 1976, 31.
47. **Burkitt, D. P., Walker, A. R. P., and Painter, N. S.,** Dietary fiber and disease, *JAMA,* 229, 1068, 1974.
48. **Mitchell, W. D. and Eastwood, M. A.,** Dietary fiber and colon function, in *Fiber in Human Nutrition,* Spiller, G. A. and Amen, R. J., Eds., Plenum Press, New York, 1976, 185.
49. **Kirwan, W. O., Smith, A. N., McConnell, A. A., Mitchell, W. D., and Eastwood, M. A.,** Action of different bran preparations on colonic function, *Br. Med. J.,* 4, 187, 1974.
50. Dietary fibre, transit-time, fecal bacteria, steroids and colon cancer in two Scandinavian populations. Report from the International Agency for Research on Cancer Intestinal Microecology Group, *Lancet,* 2, 207, 1977.
51. **Mendeloff, A. I.,** Dietary fiber, in *Present Knowledge in Nutrition,* 4th ed., Nutrition Foundation, New York, 1976, 392.
52. **Jukes, T. H.,** Food additives, *N. Engl. J. Med.,* 297, 427, 1977.
53. **Sapeike, N.,** *Food Pharmacology,* Charles C Thomas, Springfield, Ill., 1969, 108.
54. Sixth Report of FAO/WHO Expert Committee on Food Additives, WHO Tech. Rep. Ser. No. 228, World Health Organization, Washington, D.C., 1962.
55. **Hook, E. B.,** Dietary cravings and aversions during pregnancy, *Am. J. Clin. Nutr.,* 31, 1355, 1978.
56. **Edwards, C. H., McSwain, H., and Haire, S.,** Odd dietary practices of women, *J. Am. Diet. Assoc.,* 30, 976, 1954.
57. **Halsted, J. A.,** Geophagia in man: its nature and nutritional effects, *Am. J. Clin. Nutr.,* 21, 1384, 1968.
58. **Pitkin, R. M.,** Nutrition during pregnancy: the clinical approach, in *Nutritional Disorders of American Women,* Winick, M., Ed., John Wiley & Sons, New York, 1977, 27.
59. **Chanarin, I., Rothman, D., Perry, J., and Stratfull, D.,** Normal dietary folate, iron and protein intake, with particular reference to pregnancy, *Br. Med. J.,* 2, 394, 1968.
60. **Wallace, H. M. et al.,** A study of services and needs of teenage pregnant girls in the large cities of the United States, *Am. J. Public Health,* 63, 5, 1973.
61. Relation of nutrition to pregnancy in adolescence, in *Maternal Nutrition and the Course of Pregnancy,* National Academy of Science, Washington, D.C., 1970.
62. **Morris, N. M. et al.,** Reduction of low birth weight birth rates by the prevention of unwanted pregnancies, *Am. J. Public Health,* 63, 935, 1973.
63. **Terris, M. and Glasser, M.,** A life table analysis of the relation of prenatal care to prematurity, *Am. J. Public Health,* 64, 869, 1974.
64. **Davidson, S., Passmore, R., and Brock, J. F., Eds.,** *Human Nutrition and Dietetics,* 5th ed., Longman, New York, 1973, 527.
65. **Ammann, A. J. and Stiehm, E. R.,** Immunoglobulin levels in colostrum and breast milk, and serum from formula and breast-fed newborns, *Proc. Soc. Exp. Biol. Med.,* 122, 1098, 1966.
66. **Macy, I. G. and Kelly, H. J.,** Human milk and cow's milk in infant nutrition, in *Milk: The Mammary Gland and Its Secretions,* Vol. 2, Kon, S. K. and Cowie, A. T., Eds., Academic Press, New York, 1961, 265.

67. Morrison, S. D., *Human Milk: Yields, Proximate Principles and Inorganic Constituents,* Commonwealth Agricultural Bureaux, Farnham Royal, Slough, Bucks, England, 1952.

68. Fomon, S. J. and Filer, L. J., Jr., Milk and formulas, in *Infant Nutrition,* 2nd ed., Fomon, S. J., Ed., W. B. Saunders, Philadelphia, 1974, 365, 384.

69. Latham, M. C., Infant feeding in national and international perspective: an examination of the decline in human lactation, and the modern crisis in infant and young child feeding practices, *Ann. N.Y. Acad. Sci.,* 300, 197, 1977.

70. Catz, T. S. and Giacoia, G. P., Drugs and breast milk, *Pediatr. Clin. N. Am.,* 19, 151, 1972.

71. Committee on Nutrition, American Academy of Pediatrics, Commentary on breast-feeding and infant formulas, including proposed standards for formulas, *Pediatrics,* 57, 278, 1976.

72. Fomon, S. J., Ziegler, E. E., and O'Donnell, A. M., Infant feeding in health and disease, in *Infant Nutrition,* 2nd ed., Fomon, S. J., Ed., W. B. Saunders, Philadelphia, 1974, 472.

73. Hansen, A. E., Stewart, R. A., Hughes, G., and Soderhjelm, L., The relation of linoleic acid to infant feedings, *Acta Paediatr. Scand. Suppl.,* 51, 137, 1962.

74. Committee on Nutrition, American Academy of Pediatrics, Proposed changes in Food and Drug Administration regulations concerning products and vitamin-mineral supplements for infants, *Pediatrics,* 40, 916, 1967.

75. Committee on Nutrition, American Academy of Pediatrics. Iron-fortified formulas, *Pediatrics,* 47, 786, 1971.

76. Isselbacher, K. J., Galactose metabolism and galactosemia, *Am. J. Med.,* 26, 715, 1959.

77. Senior, J. R., Ed., *Medium Chain Triglycerides,* University of Pennsylvania Press, Philadelphia, 1968.

78. *Physicians' Desk Reference,* Medical Economics, Oradell, N.J., 1976, 1018.

79. Ming, H. C., Parenteral nutrition: principles, nutrient requirements, techniques and clinical applications, in *Nutritional Support of Medical Practice,* Schneider, H. A., Anderson, C. E., and Coursin, D. B., Eds., Harper & Row, New York, 1977, 152.

80. Shills, M. E., A program for total parenteral nutrition at home, *Am. J. Clin. Nutr.,* 28, 1429, 1972.

81. Shills, M. E., Guidelines for total parenteral nutrition, *J. Am. Med. Assoc.,* 220, 1721, 1972.

82. Heird, W. C., Driscoll, J. M., Jr., Schullinger, J. N., Grebin, B., and Winters, R. W., Intravenous alimentation in pediatric patients, *J. Pediatr.,* 80, 351, 1972.

83. Heird, W. C., Driscoll, J. M., Jr., and Winters, R. W., Total parenteral nutrition in very low birth weight infants, in *Size at Birth,* CIBA Symp., Elsevier-Excerpta Medica, Amsterdam, 1974, 329.

84. Heird, W. C. and Winters, R. W., Total parenteral nutrition: the state of the art, *J. Pediatr.,* 86, 2, 1975.

85. Kang, E. S., Sollee, N. D., and Gerald, P. S., Results of treatment and termination of the diet in phenylketonuria (PKU), *Pediatrics,* 46, 881, 1970.

86. Knox, W. E., Phenylketonuria, in *The Metabolic Basis of Inherited Disease,* 3rd ed., Stanbury, J. B., Wyngaarden, J. B., and Fredrickson, D. S., Eds., McGraw-Hill, New York, 1972, 266.

87. Koch, R., Acosta, P., Donnell, G., and Lieberman, E., Nutritional therapy of galactosemia, *Clin. Pediatr.,* 4, 571, 1965.

88. Moynahan, E. J., Acrodermatitis enteropathica: a lethal inherited human zinc-deficiency disorder, *Lancet,* 2, 399, 1974.

89. Herbert, V., The five possible causes of all nutrient deficiencies illustrated by deficiencies of vitamin $B_{12}$ and folic acid, *Am. J. Clin. Nutr.,* 26, 77, 1973.

90. Roe, D. A., *Drug-induced Nutritional Deficiencies,* AVI Publishing, Westport, Conn., 1976, 70.

91. Young, C. M., et al., Cooperative nutritional status studies in the Northeast Region. III. Contributions to dietary methodology studies, *Agric. Exp. Stn. Northeast Regional Publ. 10,* University of Massachusetts, Amherst, Mass., Bull. 469, 1952.

92. Adams, C. F., Nutritive Value of American Foods, Agricultural Handbook 456, Agricultural Research Service, U.S. Department of Agriculture, Washington, D.C., 1975.

93. Burke, B. S., The dietary history as a tool in research, *J. Am. Diet. Assoc.,* 23, 1041, 1947.

94. Reshef, A. and Epstein, L. M., Reliability of a dietary questionnaire, *Am. J. Clin. Nutr.,* 25, 91, 1972.

95. Fomon, S. J., Nutritional disorders of children — screening, follow-up, prevention, BCHS, Department of Health, Education, and Welfare, Washington, D.C., 1977.

96. Buzina, R., Assessment of nutritional status and food composition surveys, *Bibl. Nutr. Dieta,* 20, 24, 1974.

97. Stoudt, H. W., Damon, A., McFarland, R., et al., Weight, height and selected body dimensions of adults, U.S. Public Health Service, Washington, D.C., 1965.

98. *Build and Blood Pressure Study,* 1959, Society of Actuaries, Chicago.

99. **Christakis, G.,** The prevalence of adult obesity, in Obesity in Perspective, Fogarty International Center Series on Preventative Medicine, Vol. 2, Part 2, Bray, G. A., Ed., Department of Health, Education, and Welfare, Publ. No. (NIH)75-708, 1975.

100. **Mayer, J. and Seltzer, C. C.,** A simple criterion of obesity, *Postgrad. Med.,* 38, 1A101, 1965.

101. **Frisancho, A. R.,** Triceps skinfold and upper arm muscle size norms for assessment of nutritional status, *Am. J. Clin. Nutr.,* 27, 1052, 1974.

102. **Garn, S. M., Smith, N. J., and Clark, D. C.,** The magnitude and implications of apparent race differences in hemoglobin values, *Am. J. Clin. Nutr.,* 28, 563, 1975.

103. **Blackburn, G. L. et al.,** Nutritional and metabolic assessment of the hospitalized patient, *J. Parenteral Enteral Nutr.,* 1, 11, 1977.

104. **Bistrian, B. R., et al.,** Cellular immunity in semi-starved states in hospitalized adults, *Am. J. Clin. Nutr.,* 28, 1148, 1975.

105. **Bennett, J. R. and Baddeley, M.,** Diseases of the alimentary system. Obesity, *Br. Med. J.,* 2, 1052, 1976.

106. **Garn, S. M., Cole, P. E., and Bailey, S. M.,** Effect of parental fatness levels on the fatness of biological and adoptive children, *Ecol. Food Nutr.,* 7, 91, 1977.

107. **Kolata, G. B.,** Editorial, Obesity, *Science,* 198, 905, 1977.

108. **Bierman, E. L., Bagdade, J. D., and Porte, D.,** Obesity and diabetes: the odd couple, *Am. J. Clin. Nutr.,* 21, 1434, 1968.

109. **Goldman, A. G., Varady, P. D., and Franklin, S. S.,** Body habitus and serum cholesterol in essential hypertension and renovascular hypertension, *JAMA,* 221, 378, 1972.

110. **Keys, A.,** Overweight and the risk of heart attack and sudden death, in Obesity in Perspective, Fogarty International Center Series on Preventative Medicine, Vol. 2, Part 2, Bray, G. A., Ed., Department of Health, Education and Welfare, Publ. No. 75-708, Washington, D.C., 1975.

111. **Walker, H. C.,** Obesity. Its complications and sequelae, *Arch. Intern. Med.,* 93, 951, 1954.

112. **Auchincloss, J. H., Sipple, J., and Gilbert, R.,** Effect of obesity on ventilatory adjustment to exercise, *J. Appl. Physiol.,* 18, 19, 1963.

113. **Henschel, A.,** Obesity as an occupational hazard, *Can. J. Public Health,* 58, 491, 1967.

114. **Roe, D. A. and Eickwort, K.,** Relationships between obesity and associated health factors with unemployment among low income women, *J. Am. Med. Women's Assoc.,* 31, 193, 1976.

115. **Bortz, W. M.,** Predictability of weight loss, *JAMA,* 204, 99, 1968.

116. **Leon, G. R.,** A behavioral approach to obesity, *Am. J. Clin. Nutr.,* 30, 785, 1977.

117. **Young, C. M.,** Some thoughts on overnutrition or obesity, *Home Econ. Quart. Review of Nutrition and Food Science,* 1, 7, 1965.

118. **Young, C. M.,** Dietary treatment of obesity, in Obesity in Perspective, Fogarty International Center Series of Preventative Medicine, Vol. 2, Part 2, Bray, G. A., Ed., Department of Health, Education, and Welfare, (NIH)75-708, 1975, 361.

119. **Solomon, N.,** Improper dieting — a health hazard, *Md. State Med. J.,* 23(42), 61, 70, 1970.

120. **Overton, M. H. and Lukert, B. P.,** *Clinical Nutrition,* Year Book Medical Publishing, Chicago, 1977, 27.

121. **Linn, R.,** *The Last Chance Diet Book,* Bantam, New York, 1977.

122. **Michiel, R. R., Sneider, J. S., Dickstein, R. A., Hayman, H., and Eich, R. H.,** Sudden death in a patient on a Liquid Protein Diet, *N. Engl. J. Med.,* 298, 1005, 1978.

123. **Sullivan, A. C.,** Agents for the treatment of obesity, *Ann. Rep. Med. Chem.,* Vol. 11, Clarke, F. H., Ed., Academic Press, New York, 1976, 200.

124. **Blundell, J. E., Latham, C. J., and Lesheim, M. B.,** Differences between anorexic actions of amphetamine and fenfluramine — possible effects on hunger and satiety, *J. Pharm. Pharmacol.,* 28, 471, 1976.

125. **Rivlin, R. S.,** Therapy of obesity with hormones, *N. Engl. J. Med.,* 292, 26, 1975.

126. **Anorexiants,** in *AMA Drug Evaluations,* 2nd ed., Publishing Sciences Group, Acton, Mass., 1973, 369.

127. **Evans, E. and Miller, D. S.,** Bulking agents in the treatment of obesity, *Nutr. Metab.,* 199, 199, 1975.

128. **Gazet, J-C., Pilkington, T. R. E., Kalucy, R. S., et al.,** Treatment of gross obesity by jejunal bypass, *Br. Med. J.,* 4, 311, 1974.

129. **Thomas Holtzbach, R., Wieland, R. G., Lieber, C. S., et al.,** Hepatic lipid in morbid obesity, *N. Engl. J. Med.,* 290, 296, 1974.

130. **Compston, J. E., Laker, M. F., Woodhead, J. S., et al.,** Bone disease after jejuno-ileal bypass for obesity, *Lancet,* 2, 1, 1978.

131. **Wynder, E. L.,** Nutrition and cancer, *Fed. Proc.,* 35, 1309, 1976.

132. **Kannel, W. B.,** Status of coronary heart disease risk factors, *J. Nutr. Ed.,* 10, 10, 1978.

133. **Dahl, L. K.**, Sodium intake of the American male. Implications on the etiology of hypertension, *Am. J. Clin. Nutr.*, 6, 1, 1958.
134. **Stamler, J., Stamler, R., Rudlinger, F., Algera, G., and Roberts, R. H.**, Hypertension screening of one million Americans. Community hypertension and evaluation clinic (CHEC) program, 1973-75, *JAMA*, 235, 2299, 1976.
135. **Dustan, H. R., Terazi, R. C., and Bravo, E. L.**, Diuretic and diet treatment of hypertension, *Arch. Intern. Med.*, 133, 1007, 1974.
136. **Coleman, D. L.**, Diabetes and obesity: thrifty mutants?, *Nutr. Rev.*, 38, 129, 1978.
137. **Cohen, A. M.**, Prevalence of diabetes among different ethnic Jewish groups in Israel, *Metabolism*, 10, 50, 1961.
138. **West, K. M.**, Diabetes in American Indians and other native populations of the New World, *Diabetes*, 23, 841, 1974.
139. **Arky, R. A.**, Environmental factors influencing the development of diabetic state, in Diabetes Mellitus, Fogarty International Center Series on Preventative Medicine, Vol. 4, Fajans, S. S., Ed., Department of Health, Education, and Welfare, (NIH)76-854, 1976.
140. **West, K. M.**, Diabetes mellitus, in *Nutritional Support of Medical Practice*, Schneider, H. A., Andersen, C. E., and Coursin, D. B., Eds., Harper & Row, Hagerstown, Md., 1977, 278.
141. **Hartles, R. L. and Leach, S. A.**, Effect of diet on dental caries, *Br. Med. Bull.*, 31, 137, 1975.
142. **Scherf, H. W.**, Dental caries: prospects for prevention, *Science*, 173, 1199, 1971.
143. **Stephenson, P. E.**, Physiologic and psychotropic effects of caffeine on man, *J. Am. Diet. Assoc.*, 71, 240, 1977.
144. Anxiety symptoms and coffee drinking (editorial), *Br. Med. J.*, 1, 296, 1975.
145. **Roe, D. A.**, Physical rehabilitation and employment of AFDC recipients. Final Report, U.S. Department of Labor, Washington, D.C., 1978.
146. **Drunkenmölle, R., et al.**, Gastric acid secretion and mucosal morphology in rosacea, *Dermatologica*, 141, 385, 1970.
147. **Marks, R., Beard, R. J., Clark, M. L., et al.**, Gastrointestinal observations in rosacea, *Lancet*, 1, 739, 1967.
148. **Cohen, S. and Booth, G. H., Jr.**, Gastric acid secretion and lower-esophageal-sphincter pressure in response to coffee and caffeine, *N. Engl. J. Med.*, 293, 897, 1975.
149. **Davenport, H. W.**, *Physiology of the Digestive Tract*, 2nd ed., Year Book Medical Publishing, Chicago, Ill., 1966, 216.
150. **Mendeloff, A. I.**, Dietary fiber and health, *N. Engl. J. Med.*, 297, 811, 1977.
151. **Jukes, T. H.**, Megavitamin therapy, *JAMA*, 233, 550, 1975.
152. **Shaywitz, B. A., Siegel, N. J., and Rearson, H. A.**, Megavitamins for mineral brain dysfunction. A potentially dangerous therapy, *JAMA*, 238, 1749, 1977.
153. **Vaesrub, S.**, Vitamin abuse, *JAMA*, 238, 1762, 1977.
154. **Arnrich, L.**, Toxic effects of megadoses of fat-soluble vitamins, in *Nutrition and Drug Interrelations*, Hathcock, J. N. and Coon, J., Eds., Academic Press, New York, 1978, 751.
155. **Roe, D. A.**, Nutrient toxicity with excessive intake. II. Mineral overload, *N.Y. State J. Med.*, 66, 1233, 1966.
156. **Jacobs, J., Greene, H. and Gendel, B. R.**, Acute iron intoxication, *N. Engl. J. Med.*, 273, 1124, 1965.
157. **MacDonald, R. A.**, Hemochromatosis, *Am. J. Clin. Nutr.*, 23, 592, 1970.
158. **Perman, G.**, Hemochromatosis and red wine, *Acta Med. Scand.*, 182, 281, 1967.
159. **Seftel, H. C., et al.**, Osteoporosis, scurvy and siderosis in Johannesburg Bantu, *Br. Med. J.*, 1, 642, 1966.
160. **Singh, A., Jolly, S. S., Bansal, B. C., and Mathur, C. C.**, Endemic fluorosis. Epidemiological, clinical and biochemical study of chronic fluorine intoxication in Punjab (India), *Medicine*, 42, 229, 1963.
161. **Sass-Kortsak, A.**, Copper metabolism, in *Advances in Clinical Chemistry*, Vol. 8, Sobotka, H. and Stewart, C. P., Eds., Academic Press, New York, 1965, 39.
162. **Walkiw, O. and Doublas, D. E.**, Health food supplements prepared from kelp — a source of elevated urinary arsenic, *Clin. Toxicol.*, 8, 325, 1975.
163. **Crosby, W. H.**, Lead-contaminated health food: the tip of an iceberg, *JAMA*, 238, 1544, 1977.
164. **Lowenstein, F. W.**, Major nutrition-related risk factors in American women, in *Nutritional Disorders of American Women*, Winick, M., Ed., John Wiley & Sons, New York, 1977, 157.
165. Ten-State Nutrition Survey 1968—70. III. Clinical, anthropometry, dental, Public Health Service, Department of Health, Education, and Welfare, Publ. No. (HSM)72-8132, Atlanta, Ga., 1972.
166. Preliminary Findings of the First Health and Nutrition Examination Survey, United States, 1971—72. Dietary intake and biochemical findings, Public Health Service, Department of Health, Education, and Welfare, Publ. No. (HRA)74-1219-1, 1974.

167. **Kramer, L., Osis, D., Wiatrowski, E., and Spencer, H.,** Dietary fluoride in different areas of the United States, *Am. J. Clin. Nutr.,* 27, 590, 1974.
168. **Finch, C. A.,** Iron metabolism, in *Present Knowledge in Nutrition,* 4th ed., Nutrition Foundation, New York, 1976, 280.
169. **King, J. C., Cohenour, S. H., Calloway, D. H., and Jacobson, H. N.,** Assessment of nutritional status of teenage pregnant girls. I. Nutrient intake and pregnancy, *Am. J. Clin. Nutr.,* 25, 916, 1972.
170. **Stamp, T. C. R., Round, J. M., Rowe, D. J. F., and Haddad, J. G.,** Plasma levels and therapeutic effect of 25-hydroxycholecalciferol in epileptic patients, *Br. Med. J.,* 4, 9, 1972.
171. **Exton-Smith, A. N.,** Nutritional deficiencies in the elderly, in *Nutritional Deficiencies in Modern Society,* Howard, A. N. and McLean Baird, I., Eds., Newman Books, London, 1973, 73.
172. **Butterworth, C. E. and Blackburn, G. L.,** Hospital malnutrition and how to assess the nutritional status of a patient, *Nutr. Today,* 10, 2, 8, 1975.
173. **Davidson, S., Passmore, R., and Brock, J. F.,** Eds., Nutrition in relation to injury, surgery and convalescence, in *Human Nutrition and Dietetics,* 4th ed., Longman, New York, 1973, 446.
174. **Pearson, E. and Soroff, H. S.,** Burns, in *Nutritional Support of Medical Practice,* Schneider, H. A., Andersen, C. E., and Coursin, D. B., Eds., Harper & Row, New York, 1977, 222.
175. *Calorie Deficiencies and Protein Deficiencies,* McCance, R. A. and Widdowson, E. M., Eds., J. & A. Churchill, London, 1968.
176. **Shuster, S. and Marks, J.,** *Systemic Effects of Skin Disease,* Heinemann Med. Books, London, 1970, 54.
177. **Terepka, A. R. and Waterhouse, C.,** Metabolic observations during the forced-feeding of patients with cancer, *Am. J. Med.,* 20, 225, 1956.
178. **Gabuzda, G. J.,** Nutrition and liver disease, *Med. Clin. North Am.,* 54, 1455, 1970.
179. **Leevy, C. M., Carli, L., Frank, O., Gellene, R., and Baker, H.,** Incidence and significance of hypovitaminemia in a randomly selected municipal hospital population, *Am. J. Clin. Nutr.,* 17, 259, 1965.
180. **Barlow, A. L.,** Options for nutrition in the critically ill, *J. Am. Med. Women's Assoc.,* 33, 293, 1978.
181. **Taylor, K. B.,** Clinical applications of a chemical diet, in *Progress in Human Nutrition,* Vol. 1, Margen, S., Ed., AVI Publishing, Westport, Conn., 1971, 192.
182. **Glotzer, D. J., Boyle, P. L., and Silen, W.,** Preoperative preparation of the colon with an elemental diet, *Surgery,* 74, 703, 1973.
183. **MacFadyen, B. V., Coplan, D. E. M., III, and Dudrick, S. J.,** Surgery, in *Nutritional Support of Medical Practice,* Harper & Row, New York, 1977, 485.
184. **Shenkin, A. and Wretlind, A.,** Parenteral nutrition, *World Rev. Nutr. Diet.,* 28, 1, 1978.
185. **Smith, A. D. M.,** Veganism: a clinical survey with observations on vitamin $B_{12}$ metabolism, *Br. Med. J.,* 1, 1655, 1962.
186. **Baker, S. J., Jacob, E., Rajan, K. T., and Swaminathan, S. P.,** Vitamin $B_{12}$ deficiency in pregnancy and the peurperium, *Br. Med. J.,* 1, 1658, 1962.
187. **Hyginbottom, M. C., Sweetman, L., and Nyhan, W. L.,** A syndrome of methylmalonic aciduria, homocystinuria, megaloblastic anemia and neurological abnormalities in a vitamin $B_{12}$ deficient breast-fed infant of a strict vegetarian, *N. Engl. J. Med.,* 299, 317, 1978.
188. **Criep, L. H.,** Diagnosis of atopy, in *Dermatologic Allergy: Immunology, Diagnosis, Management,* W. B. Saunders, Philadelphia, 1967, 264.
189. **Buchman, E., Kaung, D. T., and Knapp, R. N.,** Dietary treatment in duodenal ulcer, *Am. J. Clin. Nutr.,* 22, 1536, 1969.
190. **Skala, I. and Lamacova, V.,** Diets in lactose intolerance, *Nutr. Metab.,* 13, 200, 1971.
191. **Stephenson, L. and Latham, M. C.,** Lactose tolerance tests as a predictor of milk tolerance, *Am. J. Clin. Nutr.,* 28, 86, 1975.
192. **Whay Kooh, S., Fraser, D., Reilly, B. J., et al.,** Rickets due to calcium deficiency, *N. Engl. J. Med.,* 2197, 1264, 1977.
193. **Lubchenco, L. O., Hansman, C., and Backstrom, L.,** Factors influencing fetal growth, in *Aspects of Praematurity and Dysmaturity. Nutricia Symposium,* Jonxis, J. H. P., Visser, H. K. A., and Troelstra, J. A., Eds., Stenfert Kroese, New York, 1968.
194. **Wilson, J. G.,** Environmental factors: teratogenic drugs, in *Prevention of Embryonic, Fetal and Perinatal Disease,* Brent, R. L. and Harris, M. I., Eds., Department of Health, Education, and Welfare, Publ. No. NIH 76-863, 1976.
195. **Meyer, M. B.,** How does maternal smoking affect birth weight and maternal weight gain?, *Am. J. Obstet. Gynecol.,* 131, 888, 1978.
196. **Levvis, R., Charles, M., and Patwary, K. M.,** Relationships between birth weight and selected social, environmental and medical care factors, *Am. J. Public Health,* 63, 973, 1973.
197. **Efiong, E. I. and Banjoko, M. O.,** The obstetrical performance of Nigerian primigravidae aged 16 and under, *Br. J. Obstet. Gynaecol.,* 82, 228, 1975.

198. Gatenby, P. B. B., The anaemias in Dublin, *Proc. Nutr. Soc.*, 15, 115, 1956.
199. Haworth, J. C. and Vidyasagar, D., Hypoglycemia in the newborn, *Clin. Obstet. Gynecol.*, 14, 821, 1971.
200. Hashim, S. A. and Asfeur, R. H., Tocopherol in infants fed diets rich in polyunsaturated fatty acids, *Am. J. Clin. Nutr.*, 21, 75, 1968.
201. Keenan, W. J., Jewitt, T., and Glueck, H. I., Role of feeding and vitamin K in hypoprothrombinemia of the newborn, *Am. J. Dis. Child*, 121.
202. Fomon, S. J., *Infant Nutrition*, 2nd ed., W. B. Saunders, Philadelphia, 1974, 118.
203. Babson, S. G., Feeding the low birth weight infant, *J. Pediatr.*, 79, 694, 1971.
204. Roe, D. A., *Alcohol and the Diet*, AVI Publishing, Westport, Conn., 1979.
205. Hillman, R. W., Alcoholism and malnutrition, in *The Biology of Alcoholism*, Vol. 3, Kissin, B. and Begleiter, H., Eds., Plenum Press, New York, 1974, 513.
206. Saville, P. D., Alcohol-related skeletal disorders, *Ann. N.Y. Acad. Sci.*, 252, 287, 1975.
207. Eichner, E. R. and Hillman, R. S., The evolution of anemia in alcoholic patients, *Am. J. Med.*, 50, 218, 1971.
208. Lieber, C. S., Liver disease and alcohol: fatty liver, alcoholic hepatitis, cirrhosis and their interrelationships, *Ann. N. Y. Acad. Sci.*, 252, 63, 1975.
209. Mezey, E., Jow, E., Slavin, R. E., et al., Pancreatic function and intestinal absorption in chronic alcoholism, *Gastroenterology*, 59, 657, 1970.
210. Victor, M., The role of nutrition in the alcoholic neurologic diseases, *J. Clin. Invest.*, 39, 1037, 1960.
211. Warin, R. P., Food factors in urticaria, *J. Hum. Nutr.*, 30, 179, 1976.
212. Noid, H. E. and Winkelman, N. R. K., Diet plan for patients with salicylate-induced urticaria, *Arch. Dermatol.*, 109, 866, 1974.
213. Michaelsson, G., Pettersson, L., and Juhlin, L., Purpura caused by food and drug additives, *Arch. Dermatol.*, 109, 49, 1974.
214. Vought, R. L., Brown, F. A., and Wolff, J., Erythrosine: an adventitious source of iodide, *J. Clin. Endocrinol. Metab.*, 34, 747, 1972.
215. Adams, R. M., *Occupational Contact Dermatitis*, Lippincott, New York, 1969, 145, 225.
216. Erikson, K. E., Allergy to ethylene diamine, *Arch. Dermatol.*, 111, 791, 1975.
217. Golbert, T. M., Food allergy and immunologic disease of the gastrointestinal tract, in *Allergic Diseases. Diagnosis and Management*, Patterson, R., Ed., Lippincott, New York, 1972, 355.
218. Asatoor, A. M., Levi, A. J., and Milne, D. D., Tranylcypromine and cheese, *Lancet*, 2, 733, 1963.
219. Horwitz, D., Lovenberg, W., Engelman, K., and Sjoerdsma, A., Monamine oxidase inhibitors, tyramine, and cheese, *JAMA*, 188, 1108, 1964.
220. Nuessle, W. F. and Norman, F. C., Pickled herring and tranylcypromine reaction, *JAMA*, 192, 726, 1965.
221. Thomas, J. C. S., Monamine oxidase inhibitors in cheese, *Br. Med. J.*, 2, 1406, 1963.
222. Tyrer, P., Candy, J., and Kelly, G., A study of the clinical effects of phenelzine and placebo in the treatment of phobic anxiety, *Psychopharmacologia*, 32, 237, 1973.
223. Spivak, S. D., Procarbazine, *Ann. Intern. Med.*, 81, 795, 1974.
224. Morgan, R. and Cagan, E. J., Acute alcohol intoxication, the disulfiram reaction and methyl alcohol intoxication, in *The Biology of Alcoholism*, Vol. 3, Kissin, B. and Begleiter, H., Eds., Plenum Press, New York, 1974, 171.
225. Asmussen, E., Hald, J., and Larson, V., The pharmacological action of acetaldehyde on the human organism, *Acta Pharm. Toxicol.*, 4, 311, 1948.
226. Griffin, J. P. and D'Arcy, P. F., *A Manual of Adverse Drug Reactions*, John Wright & Sons, Bristol, 1975, 60.
227. Martindale, *The Extra Pharmacopoeia*, 26th ed., Blacow, N. W., Ed., Pharen Press, London, 1972, 157.
228. Bergman, H., Hypoglycemia coma during sulfonylurea therapy, *Acta Med. Scand.*, 177, 287, 1965.
229. Gross, E. G., Dexter, J. D., and Roth, R. G., Hypokalemic myopathy with myoglobinuria associated with licorice ingestion, *N. Engl. J. Med.*, 274, 602, 1966.
230. Roussak, N. J., Fetal hypokalemic alkalosis with tetany during licorice and PAS therapy, *Br. Med. J.*, 1, 360, 1952.
231. Welling, P. G., Influence of food and diet on gastrointestinal drug absorption: a review, *J. Pharmacokinet. Biopharm.*, 5, 291, 1977.
232. Greenblatt, D. J., Duhme, D. W., Koch-Weser, J., and Smith, T. W., Bioavailability of digoxin tablets and elixir in the fasting and postprandial states, *Clin. Pharm. Ther.*, 16, 444, 1974.
233. Welling, P. B., Lyons, L. L., Craig, W. A., and Trochta, G. A., Influence of diet and fluid on bioavailability of theophylline, *Clin. Pharm. Ther.*, 17, 475, 1975.
234. Roe, D. A., Diet-drug interactions and incompatibilities, in *Nutrition and Drug Interrelations*, Academic Press, New York, 1978, 319.

235. Greenberger, N. J., Effects of antibiotics and other agents on the intestinal transport of iron, *Am. J. Clin. Nutr.*, 26, 104, 1973.
236. Gill, G. V., Rifampicin and breakfast, *Lancet*, 2, 1135, 1976.
237. Melander, A., Danielson, K., Schersten, B., and Wahlin, E., Enhancement of the bioavailability of propranolol and metoprolol by food, *Clin. Pharm. Ther.*, 22, 108, 1977.
238. Melander, A., Danielson, K., Schersten, B., Thulin, T., and Wahlin, E., Enhancement by food of canrenone bioavailability from spironolactone, *Clin. Pharm. Ther.*, 22, 100, 1977.
239. Melander, A., Danielson, K., Hanson, A., Roudell, B., Schersten, B., Thulin, T., and Wahlin, E., Enhancement of hydralazine bioavailability by food, *Clin. Pharm. Ther.*, 22, 104, 1977.
240. Crounse, R. G., Human pharmacology of griseofulvin: the effect of fat intake on gastrointestinal absorption, *J. Invest. Dermatol.*, 37, 529, 1961.
241. Roe, D. A., Effects of drugs on nutrition, *Life Sci.*, 15, 1219, 1974.
242. Fleisher, N., et al., Chronic laxative-induced hyperaldosteronism and hypokalemia simulating Bartters syndrome, *Ann. Int. Med.*, 70, 791, 1969.
243. Joong, S., et al., Hypokalemia vacuolar myopathy associated with chlorthalidone treatment, *JAMA*, 216, 1858, 1971.
244. Hansten, P. D., *Drug Interactions*, 2nd ed., Lea & Febiger, Philadelphia, 1973, 357.
245. Haynes, R. C. and Larner, J., Adrenocorticotropic hormone; adrenocortical steroids and their synthetic analogs, inhibitors of adrenocortical steroid biosynthesis, in *The Pharmacological Basis of Therapeutics*, 5th ed., Goodman, L. S. and Gillman, A., Eds., Macmillan, New York, 1975, 1485.
246. Sullivan, J. F. and Lankford, H. G., Zinc metabolism and chronic alcoholism, *Am. J. Clin. Nutr.*, 17, 57, 1965.
247. Lim, P. and Jacob, E., Magnesium deficiency in patients on long-term diuretic therapy for heart failure, *Br. Med. J.*, 3, 620, 1972.
248. Wester, P. O., Zinc during diuretic treatment, *Lancet*, 1, 578, 1975.
249. D'Arcy, P. F. and Griffin, J. P., *Iatrogenic Diseases*, Oxford University Press, New York, 1972, 105.
250. Dennis, N. R., Rickets following anticonvulsant therapy, *Proc. R. Soc. Med.*, 65, 730, 1972.
251. Greenlaw, R., Pryor, J., Winnacker, J., et al., Osteomalacia from anticonvulsant drugs — therapeutic implications, *Clin. Res.*, 20, 56, 1972.
252. Lotz, M., Zisman, E., and Bartter, F. C., Evidence for a phosphorus depletion syndrome in man, *N. Engl. J. Med.*, 278, 409, 1968.
253. Bloom, W. L. and Flinchum, D., Osteomalacia with pseudofractures caused by the ingestion of aluminum hydroxide, *JAMA*, 174, 1327, 1960.
254. Richens, A., *Drug Treatment of Epilepsy*, H. Kimpton Publishers, London, 1976, 127.
255. Roe, D. A., Nutrition and the contraceptive pill, in *Nutritional Disorders of American Women*, Winick, M., Ed., John Wiley & Sons, New York, 1977, 37.
256. Kappas, A., Anderson, K. E., Conney, A. H., and Alvares, A. P., Influence of dietary protein and carbohydrate on antipyrine and theophylline metabolism in man, *Clin. Pharm. Ther.*, 20, 643, 1976.
257. Lewis, G. P., Jusko, W. J., Burke, C. W., and Graves, L., Prednisone side effects and serum protein levels, *Lancet*, 2, 778, 1971.
258. Mehta, S., Kalsi, H. K., Jayaraman, S., and Mathur, V. S., Chloramphenicol metabolism in children with protein-calorie malnutrition, *Am. J. Clin. Nutr.*, 28, 977, 1975.
259. Zannoni, V. G. and Sato, P. H., Effects of ascorbic acid on microsomal drug metabolism, *Ann. N.Y. Acad. Sci.*, 258, 119, 1975.
260. Hollingsworth, D. F., Nutrition education, *J. Hum. Nutr.*, 32, 209, 1978.
261. Richmond, J. B., Hearings Before the Select Committee On Nutrition and Human Needs of the United States Senate, 95th Congress, 1st Session. Diet related to killer diseases, VIII. U.S. Government Printing Office, Washington, D. C., 1977, 44.
262. Promoting Optimal Nutritional Health of the Population in the United States, Policy Statement of The American Dietetic Association, *J. Am. Dent. Assoc.*, 55, 452, 1969.
263. Position Paper on the Nutrition Component of Health Services Delivery Systems, *J. Am. Dent. Assoc.*, 58, 538, 1971.
264. Egan, M. C. and Hallstrom, B. J., Building nutrition services in comprehensive health care, *J. Am. Diet. Assoc.*, 61, 491, 1972.

# INDEX

## A

# Q

# R

For Product Safety Concerns and Information please contact our EU
representative  GPSR@taylorandfrancis.com
Taylor & Francis Verlag GmbH, Kaufingerstraße 24, 80331 München, Germany